PARAMECIUM
Genetics
and Epigenetics

PARAMECIUM
GENETICS
AND EPIGENETICS

WITHDRAWN

G. H. BEALE
JOHN R. PREER, JR.

CRC Press is an imprint of the
Taylor & Francis Group, an **informa** business

CRC Press
Taylor & Francis Group
6000 Broken Sound Parkway NW, Suite 300
Boca Raton, FL 33487-2742

© 2008 by Taylor & Francis Group, LLC
CRC Press is an imprint of Taylor & Francis Group, an Informa business

No claim to original U.S. Government works
Printed in the United States of America on acid-free paper
10 9 8 7 6 5 4 3 2 1

International Standard Book Number-13: 978-0-415-25785-5 (Hardcover)

This book contains information obtained from authentic and highly regarded sources Reasonable efforts have been made to publish reliable data and information, but the author and publisher cannot assume responsibility for the validity of all materials or the consequences of their use. The Authors and Publishers have attempted to trace the copyright holders of all material reproduced in this publication and apologize to copyright holders if permission to publish in this form has not been obtained. If any copyright material has not been acknowledged please write and let us know so we may rectify in any future reprint

Except as permitted under U.S. Copyright Law, no part of this book may be reprinted, reproduced, transmitted, or utilized in any form by any electronic, mechanical, or other means, now known or hereafter invented, including photocopying, microfilming, and recording, or in any information storage or retrieval system, without written permission from the publishers.

For permission to photocopy or use material electronically from this work, please access www.copyright.com (http://www.copyright.com/) or contact the Copyright Clearance Center, Inc. (CCC) 222 Rosewood Drive, Danvers, MA 01923, 978-750-8400. CCC is a not-for-profit organization that provides licenses and registration for a variety of users. For organizations that have been granted a photocopy license by the CCC, a separate system of payment has been arranged.

Trademark Notice: Product or corporate names may be trademarks or registered trademarks, and are used only for identification and explanation without intent to infringe.

Library of Congress Cataloging-in-Publication Data

Beale, Geoffrey.
  Paramecium: Genetics and Epigenetics / Geoffrey Beale and John R. Preer.
    p. ; cm.
  Includes bibliographical references and index.
  ISBN 978-0-415-25785-5 (hardcover : alk. paper) 1. Paramecium--Genetics. I. Preer, John R. II. Title.
  [DNLM: 1. Paramecium--genetics. 2. Epigenesis, Genetic. QX 151 B366p 2008]

QL368.H87B43 2008
572.8'29495--dc22   2007048391

Visit the Taylor & Francis Web site at
http://www.taylorandfrancis.com

and the CRC Press Web site at
http://www.crcpress.com

Dedication

This book is dedicated to all the people who have worked on Paramecium *and tried to get it right.*

Contents

Foreword xiii
Acknowledgments... xvii
About the Authors... xix

Chapter 1 Introduction.. 1

1.1   Introduction... 1
1.2   Genetics... 1
1.3   *Paramecium* 2
1.4   The Early Workers 3
1.5   Discarded Hypotheses.. 5
1.6   Epigenetics 6
1.7   Conclusions 6
References.. 7

Chapter 2 Mating Types in *Paramecium* 11

2.1   Mating Types and the Species Problem in Ciliates................................. 13
References.. 17

Chapter 3 General Description of the Protista and of *Paramecium* 19

3.1   Introduction... 19
3.2   Morphology... 21
        3.2.1   The Cortex 21
        3.2.2   The Cytoplasm 24
        3.2.3   The Nuclei... 24
3.3   Reproduction 29
        3.3.1   Asexual Fission.. 29
        3.3.2   Conjugation 30
        3.3.3   Autogamy 34
        3.3.4   Macronuclear Regeneration 35
        3.3.5   Cytogamy 36
        3.3.6   Hemixis.. 36
3.4   The Life Cycle... 36
References.. 39

**Chapter 4 Methods of Studying Genetic Processes in Organisms of the
                     Paramecium aurelia Species Group**... 43

4.1   Isolation of Cells and Preparation of Pure Clones 43
4.2   Obtaining Complementary Mating Types 44

4.3	Mendelian Transmission	44
4.4	Caryonidal Transmission	45
4.5	Cytoplasmic Transmission	46
4.6	Methods of Identifying Genetically Defined Genes with Their DNA	48
	4.6.1 Identification of the RNAs Produced by Genes	48
	4.6.2 Identification Based on Protein Sequence	49
	4.6.3 Identification Based on Similarity of Proteins	49
	4.6.4 Functional Complementation	49
	4.6.5 MALDI-TOF Analysis of *Paramecium* Proteins	49
References		50

Chapter 5   The Determination of Mating Types in *Paramecium* 51

5.1	Introduction	51
5.2	Caryonidal Inheritance	51
5.3	The Cytoplasmic Effect	53
5.4	Independence of the Cytoplasmic Effect and the Phenotype	54
5.5	The O* Phenotype: A New Cytoplasmic State	56
5.6	The Pure E Mutant	56
5.7	Micronuclear Determination	57
5.8	Selfers	57
5.9	Mating Types in Species Other Than Those in the *P. aurelia* Complex	58
	5.9.1 *P. bursaria*	58
	5.9.2 *P. caudatum*	58
	5.9.3 *P. multimicronucleatum*	59
5.10	A Hypothesis to Explain Mating Type Inheritance	60
References		60

Chapter 6   Symbionts and Mitochondria of *Paramecium* 63

6.1	Introduction	63
6.2	The Plasmagene Hypothesis	64
6.3	The Kinds of Symbionts	68
	6.3.1 The First Killers Found in *P. biaurelia*	70
	6.3.2 The Killers Found in *P. tetraurelia*	73
	6.3.3 The Mate Killers	76
	6.3.4 The Rapid Lysis Killers	77
	6.3.5 Mutants	78
	6.3.6 *Tectobacter:* The Swimmers	79
	6.3.7 A Nuclear Symbiont	79
	6.3.8 A Nuclear Symbiont in One Species and a Cytoplasmic in Another	82
6.4	The Metagon	82
6.5	Mitochondria	86
	6.5.1 Introduction	86
	6.5.2 Genetics of Mitochondrial Characters in *Paramecium*	88

Contents

	6.5.3	The Mitochondrial Genome in *Paramecium* 91
	6.5.4	The Mitochondrial Genetic Code 92
6.6	Conclusions 92
References 93

Chapter 7 Determination of i-Antigens 99

- 7.1 Introduction 99
- 7.2 The Immobilization Test and Serotypes 100
- 7.3 The Genetic System of the i-Antigens 103
  - 7.3.1 Early Work by Sonneborn: Genes, Cytoplasm, and Environment 103
  - 7.3.2 Environmental Control of the Cytoplasm 104
  - 7.3.3 Allelic Variants of Antigen-Determining Genes 110
  - 7.3.4 Control of the Cytoplasmic State by the i-Antigen Genes ... 113
  - 7.3.5 Heterozygotes and Dominance 115
- 7.4 Chemistry and Molecular Biology of the i-Antigens 115
  - 7.4.1 Isolation of the i-Antigens 115
  - 7.4.2 The mRNAs for the i-Antigen Genes 117
  - 7.4.3 Sequencing of the i-Antigen Genes 117
  - 7.4.4 The Unusual Genetic Code of Ciliates 119
  - 7.4.5 The I-Antigen Gene Sequences of *Paramecium* 120
    - 7.4.5.1 Periodicity 120
    - 7.4.5.2 Tandem Repeats 122
    - 7.4.5.3 Allelic and Nonallelic Variations 122
    - 7.4.5.4 Start and Stop Codons 123
    - 7.4.5.5 Isogenes 124
- 7.5 Control of Expression of i-Antigen Genes 124
  - 7.5.1 Hypotheses to Explain Control 124
  - 7.5.2 Positional Control 126
  - 7.5.3 Early Observations on Serotype Switching 127
  - 7.5.4 Importance of Gene Structure 128
  - 7.5.5 Run-on Transcription 128
  - 7.5.6 Promoters of i-Antigen Genes 129
  - 7.5.7 The Mutual Exclusion Test 131
- 7.6 Conclusions 132
- References 132

Chapter 8 Micronuclei and Macronuclei 139

- 8.1 Ploidy Levels 139
- 8.2 Activity of Micronuclei and Macronuclei 139
- 8.3 Programmed Chromosomal Breaks and Telomeres 140
- 8.4 Internal Eliminated Sequences 141
- 8.5 Induced Mendelian Deletions 143
- 8.6 Maintaining Genic Balance at Macronuclear Division 143

8.7	The *Paramecium* Genome Project	145
8.8	Nuclear Dualism	145
References		146

Chapter 9 Ribosomal RNA and DNA 151

9.1	The Early Work of Findly and Gall	151
9.2	More Recent Studies	152
9.3	Micronuclear DNA and a Model	154
9.4	Variants	154
References		156

Chapter 10 Cortical Morphogenesis 157

10.1	Double Animals	157
10.2	Cortical Mutants	158
10.3	Trichocysts	160
References		162

Chapter 11 Behavior 165

11.1	Introduction	165
11.2	Mutants	165
11.3	Genetics	165
11.4	Cell Models	165
11.5	Measuring the Electric Properties of the Cell Membrane	166
11.6	The Avoiding Reaction	167
11.7	Ion Channels	168
	11.7.1 Ca^{2+} Channels	168
	11.7.2 K^+ Channels	168
	11.7.3 Na^+ Channels	169
	11.7.4 Mg^{2+} Channels	169
11.8	Behavioral Responses Other Than the Avoiding Reaction	169
11.9	Conclusions	170
References		171

Chapter 12 Epigenetics 175

12.1	Symbionts and Mitochondria	175
12.2	Serotype Inheritance	176
12.3	Mating Type Inheritance	176
	12.3.1 Caryonidal Inheritance of Mating Type	176
	12.3.2 Cytoplasmic Inheritance	177
12.4	Cortical Inheritance	177
12.5	Homology-Dependent Inheritance	177
	12.5.1 d48: A Homology-Dependent Mutant	177
	12.5.2 Induction and Repair of Homology-Dependent Mutants	179

   12.5.3 Homology-Dependent Mutants That Fail to Excise IESs 181
   12.5.4 The Elimination of Foreign DNA 181
   12.5.5 Theories of PTGS 181
 12.6 Conclusions 182
References 183

Index 185

Foreword

Biology is probably one of today's most rapidly evolving sciences. A little more than half a century ago, the discovery of the structure of DNA was the starting point of a technical revolution in the field of genetics, which can be defined as the study of the mechanisms of heredity during the reproduction of living organisms. The techniques used to decipher the genetic makeup of species have been developing at a truly amazing pace ever since, and scientists are now accumulating vast amounts of molecular data, such as the sequences of complete genomes, at a speed that could not have been anticipated only a few years ago.

However, these sophisticated techniques also have a cost. Modern biology is under ever-increasing pressure to ensure the best returns on time and money invested, turning the professional lives of researchers into an endless race to exploit molecular data and publish findings as fast as possible, so as to be able to compete for the necessary funding. Since DNA-based inheritance gained wide acceptance as one of the basic principles of life, spectacular progress has been made in our understanding of the complex molecular mechanisms that govern gene replication and gene expression. But the very competitive nature of the system has probably not encouraged the emergence of fundamentally new concepts. The history of epigenetics, the study of inherited variations that cannot be attributed to DNA sequence variants, is an example. Even before the structure of DNA was solved, cases of inherited variation had been described that could not be explained by the Mendelian transmission of genes; but the rush toward more straightforward, grant-earning DNA studies hampered the search for alternative inheritance mechanisms for some time. Only recently has it been recognized that epigenetic mechanisms are just as essential for life as genome replication, and molecular tools are now quickly exploring a whole new dimension.

New concepts often spring from the study of a different model organism. The complexity of life forms is matched only by their diversity, and each species has the potential to reveal particular aspects that are not easily observed in others. It is fortunate that species of the *Paramecium aurelia* group were among the chosen few extensively studied in the pregenomic era, mostly thanks to the pioneering work of T. M. Sonneborn. The countless hours spent at the microscope in order to describe the complexity of their cellular structure and to work out the details of their life cycle would now seem to be an academic luxury that we can no longer afford, and one wonders whether it would be at all possible today to find the time, patience, and dedication needed to establish a new species as a model by developing practical methods and translating the peculiarities of its biology into questions of general significance.

As this book makes clear, *Paramecium* has much to offer as a eukaryotic model. It combines the advantages of a free-living, unicellular organism that can be grown rapidly in simple media with those of having a large cell size with an exquisitely complex structure and behavior. The *Paramecium* cell has antero-posterior, dorso-ventral, and left-right asymmetries and resembles a small animal, constantly swimming around to capture the bacteria on which it feeds through the coordinated beating of thousands of cilia. It can react to obstacles, predators, and chemical or electrical stimuli by swimming backwards, changing direction, or swimming faster. The cell contains organelles that can be likened to the organs of metazoans: a digestive system comprising a mouth, digestive vacuoles and an anal pore, two contractile vacuoles that pump water to regulate osmotic pressure, and defensive vesicles that can be discharged as sharp crystalline needles to protect the cell from its predators. These characteristics have long made it a useful model to study the morphogenesis of cellular structures during growth, regulated exocytosis, or the electrophysiological properties of a variety of ion channels located in the membrane.

Paramecium further offers the opportunity to study a real developmental process of germline/soma differentiation in a single-celled eukaryote. Indeed, one distinctive feature of ciliates is that they contain two distinct kinds of nuclei within the same cytoplasm. The diploid micronuclei are transcriptionally silent during vegetative growth and only serve germline functions: they undergo meiosis during sexual events to transmit the genome to the next sexual generation. The polyploid macronuclei, on the other hand, have purely somatic functions: while their genome is expressed and therefore governs the cell's phenotype, they are not transmitted to sexual progeny and have to be replaced in each generation. Although the new micronuclei and macronuclei both develop from copies of the single zygotic nucleus after fertilization, over the years it has become clear that the macronuclear genome is a much rearranged version of the germline genome present in the micronucleus. The *Paramecium* genome was recently shown to contain close to 40,000 genes (i.e., many more than in the human genome), and the cell can control the gene content of the developing macronucleus through alternative rearrangements of the germline genome. Thus, like multicellular organisms, *Paramecium* has the potential to produce a wide variety of differentiated cell types from a single genotype, although each of these types has to be able to live on its own.

Perhaps the most decisive advantage of *P. aurelia* as a model eukaryote lies in the peculiarities of its two modes of sexual reproduction, which can be easily managed in the lab and together make it particularly well suited for genetic analysis. Autogamy is a self-fertilization process by which a single cell gives rise to progeny with an entirely homozygous genotype, a precious help to the geneticist. Conjugation is the reciprocal fertilization of two cells of complementary mating types. In *P. aurelia* species, this always results in the two ex-conjugants having identical genotypes, whatever the genotypes of the parental cells. Furthermore, cytoplasm is not normally exchanged between the two mates. Each cross is therefore equivalent to the study of monozygotic twins that develop in different environments, in this case within different parental cells: any phenotypic difference between the two clones can then be ascribed to something other than Mendelian genes.

When Sonneborn discovered the mating types of *P. aurelia* in 1937, he was among those biologists who believed that chromosomes could not be the sole basis of heredity and that the cytoplasm had to be involved, if only to coordinate gene action. His genetic analyses of different strains did reveal conventional cases of inheritance that could be explained by the Mendelian segregation of alleles, but his attention focused on other cases, in which the phenotypic differences between the two parental cells were maintained in their respective offspring, following cytoplasmic lines rather than chromosomal genes. The first of these characters to be described in *P. tetraurelia*, the mating type itself, served as a paradigm for a growing list of non-Mendelian phenomena, among which are the *killer* trait (the capacity of some strains to kill other, sensitive strains) and the serotype (now known to be determined by the expression of a single surface protein chosen from among a multigene family, with occasional antigenic variation). Ironically, such a conspicuous occurrence of cytoplasmic inheritance probably contributed, after a brief period of glory, to the demise of *Paramecium* as a model, because this weird organism was perceived as a threat to nuclear orthodoxy.

An interesting aspect of this book is that it tells us not only about the success of *Paramecium* research, but also about its failures, and one of them is worth mentioning here. The plasmagene hypothesis was proposed by Sonneborn in the late 1940s, in an effort to unify the diverse cases of cytoplasmic inheritance into a single theoretical model. Plasmagenes were defined as cytoplasmic particles endowed with the genetic properties of reproduction and mutability, and were assumed to act in concert with cognate nuclear genes to control the phenotype. Some plasmagenes, such as those involved in serotype inheritance, appeared to originate from nuclear genes and also to depend on these genes for their maintenance in the cytoplasm. Further research, however, soon revealed that the cases studied corresponded to very different biological phenomena, and the search for a unifying theory was abandoned. The killer trait was shown by J. R. Preer Jr. to be determined by a bacterial endosymbiont living in the *Paramecium* cytoplasm. Although mating type genes still have to be identified, a possible inheritance mechanism was suggested by the discovery that the developing macronucleus can reproduce alternative genome rearrangements present in the maternal macronucleus. The cytoplasmic inheritance of serotypes, on the other hand, does not involve genome rearrangements; how the maternal macronucleus influences the developing macronucleus, across the cytoplasm, to express the same surface antigen gene remains a mystery. It should be noted that the plasmagene hypothesis may turn out to be essentially correct in this case. Indeed, the newly discovered RNA interference pathway involves the transcription of nuclear genes, the independent replication of RNA molecules by RNA-dependent RNA polymerases, and the specific action of double-stranded RNA on the expression of homologous genes: nothing could better fit the definition of plasmagenes originating from genes and maintained by them.

Although *Paramecium* had disappeared from textbooks for a while, it never died out in the labs, and its aficionados, though few, were strong enough to have its genome sequenced before it was too late. In the current postgenomic era, epigenetics has become not only fully respectable, but even one of the most exciting frontiers of molecular biology. Many new and powerful tools, such as functional complementation, reverse genetics, and genomic analyses, are now available for *Paramecium* research. This is what makes this book so valuable and its publication so timely. Both authors have spent the best part of their careers studying this organism's oddities, and their legacy is twofold. They provide younger scientists with a full account of seventy years of *Paramecium* genetics summarizing a wealth of time-consuming observations and remarkable phenomena, some of which have yet to be translated into state-of-the-art molecular models. And young scientists will also benefit from this account if they feel convinced, as I have been, that the power to rationalize the unconventional will always be more precious than the most sophisticated technical achievements of the day.

Eric Meyer

Laboratoire de Génétique Moléculaire
Ecole Normal Supérieure

Paris, France
August 10, 2007

Acknowledgments

The authors express appreciation to those who helped with this book. We are grateful to Andrew, Steven, and Duncan, the sons of G. H. Beale, for their advice and help in communicating with Geoffrey during his illness in Edinburgh. Richard Carter in Edinburgh deserves our many thanks for supplying information about Beale and his work. Our special heartfelt thanks are for David Walliker (recently deceased) for his constant encouragement and help, for commenting on much of the manuscript and acting as the primary liason between Beale and Preer, while dealing with his own serious medical problems. We thank Bruce Byrne for a cursory examination of the chapter on mating types on a visit to Bloomington. We are especially grateful to Ching Kung, who examined the chapter on behavior in some detail and made suggestions for revisions. James D. Forney also provided special unpublished DNA sequences. We are so pleased that Eric Meyer agreed to write the preface for this book, although he was at the time unusually busy with his growing laboratory. Thanks to "Bertie" Preer, John's wife, for her help with this project. As with many other projects, she has been at his side while *Paramecium* led them on a happy chase through decades of research. The authors also appreciate the special gifts of those who kept the laboratories in good running order over the years and those who joined us at some stage in their research careers. These people with their commitment to research, with their interests and lively discussions, made our laboratories agreeable workplaces. We especially appreciate Bertina Rudman, who began in the Preer laboratory at the University of Pennsylvania as an assistant in the fifties and retired in 2005 from Indiana University when it was decided to stop molecular work. We are grateful to Sidney Pollack and the late Artur Jurand who spent many summers and semesters helping with research in the Preer laboratory in Bloomington. We also acknowledge financial support in the writing of this book from Indiana University. In no way are any of the above responsible for errors in this work, for that responsibility lies with the authors.

About the Authors

G. H. Beale, MBE, was born June 11, 1913, in Wandsworth, London, son of Herbert Walter Beale and Elsie Beaton Beale. From an early age until he joined the army in 1941 he lived in Wallington, Surrey. After Sutton County School he entered Imperial College in 1931.

After four years and with excellent recommendations from Imperial College, he was hired by J. B. S. Haldane at the John Innes Horticultural Institution, a lively place to study genetics. He did some of the genetic studies for Scott-Montcrief's chemical work on the anthocyanin pigments of plants. Nominally his work there was supervised by Haldane, and when he left, by Darlington. In 1938 Beale wrote his thesis on a genetic analysis of *Verbena*, which with his oral exam satisfied the University of London examiners for the PhD degree in 1938.

He was called for military service in the Intelligence Corps in January 1941. While at the John Innes he attended evening classes in Russian, a happenstance that led to his posting in the British Army during WWII to northern Russia. He spent 1941–1943 in Archangel and Murmansk, serving as liason between the British Army and the Russians at the base. In the rank of captain he was sent to Finland to join the Allied Control Commission, 1944–1946. In 1946 he left the Army.

To get back into science he wanted to go to the United States. At that time Cold Spring Harbor was one of the very best places in the United States to study genetics—Demerec, Luria, Lederberg, Delbrück, and McClintock were there. Demerec gave Beale a job in his lab, where he determined the spontaneous mutation rate of *E. coli* from sensitivity to resistance to phage T1, and wrote a paper on it. At CSH he met Tracy Sonneborn, who supported his application for a Rockefeller fellowship to come to Indiana University in Bloomington to work with him, provided he had the assurance of employment in Great Britain at the end of the fellowship. This provision was met when Waddington offered him a job in the new genetics unit at the University of Edinburgh. He was in Bloomington for nine months, starting in 1946. It was a splendid place for genetics—Muller, Sonneborn, Luria, and Cleland were there. In 1954 he published his excellent monograph "The Genetics of *Paramecium aurelia*."

He spent many years devoted to experiments and the study of *Paramecium* genetics and to teaching at the University of Edinburgh, starting in 1948. In the early sixties, with funds in part from the Wellcome Trust, he set up the Protozoan Genetics Unit at the University of Edinburgh. For many years he and his master's, doctoral, and postdoctoral students did basic genetic *Paramecium* research that resulted in at least fifty-two publications. His main contribution to genetics was in being the first to show that antigens in the group A species of the *P. tetraurelia* complex of species were epigenetically inherited at a time when it was thought otherwise. His studies on antibody labeling revealed information about the distribution of antigens on the surface of cells. He also discovered mitochondrial inheritance in *Paramecium* and showed how mitochondria do not fuse and recombine their genes in the cytoplasm of *Paramecium* as they do in yeast. He became a fellow of the Royal Society in 1959, and of the Royal Society of Edinburgh in 1966. In 1963 he became Royal Society research professor, a position he held until his retirement. In 1978 he became honorary research professor.

In later years he turned his attention to medically important protozoans, the malaria parasites; and his laboratory was then devoted entirely to the study of these organisms. The first practical method of making genetic crosses between malaria parasites was developed in his laboratory. He interacted with many research groups on the King's Buildings campus at the University of Edinburgh, and dating from the start of his work on *Plasmodium*, interest in parasitology spread in the area. He established a relationship with malaria parasite researchers at Chulalongkorn University in Thailand that extended from the late 1970s to 2001. This group characterized up to one thousand cultured isolates of *Plasmodium falciparum*, charted the emerging drug resistance in the area, and demonstrated the genetic diversity of these human malaria parasites. He made frequent trips to Bangkok. Approximately nineteen papers resulted from this collaboration. For his work in Thailand he received an honorary degree directly from the hand of the king of Thailand. He did not abandon *Paramecium* and in later years went to FASEB meetings of the Molecular Biology of Ciliates in the United States and regularly spent a month in Preer's lab in Bloomington, keeping up his interest in *Paramecium* research. In 1998 in a letter to a friend he mused, "Sometime I must revise my *Paramecium* story," referring to his 1954 monograph, and he soon began in earnest to do so. This book, with its emphasis on how the genetics of *Paramecium* developed over the years, is in large measure the result of that endeavor.

John R. Preer Jr. is distinguished professor emeritus in the Department of Biology at Indiana University in Bloomington, where he has taught and done research on *Paramecium* since 1968. He retired in 1988, but continued working in his laboratory until 2005. He was born in Ocala, Florida, in 1918, son of John R. Preer and Ruth Williams Preer. He began his career in biology as a high school and then college student, working in the Agriculture Experiment Station, University of Florida at Gainesville. He studied an order of insects, *Thysanoptera* (thrips), and after a time became proficient in identifying and classifying thrips. He published a few papers and corresponded with thrips researchers around the globe. As a result of his work in taxonomy, two species of thrips, *Preeriella minuta* and *Haplothrips preeri*, were named for him. *Preeriella* was a new genus as well.

He expected to follow up his taxonomic studies in graduate school and was accepted at Indiana University in 1939 to work with Dr. A. C. Kinsey, who was beginning to turn his attention to human sexuality. A course with Tracy Sonneborn resulted in Preer's switch from taxonomy to *Paramecium* to learn about its genetics in Sonneborn's laboratory. His graduate work was interupted by World War II and service in the U.S. Army. As a medical clerk he went to England, then returned to this country to attend Officers' Candidate School. He was commissioned as a lieutenant and then assigned to a base in Texas, where he instructed bomber pilots in a high-altitude training unit until war's end. He returned to Indiana University in 1946. In 1947 he received his PhD degree.

About the Authors

He then accepted a position in the Department of Zoology at the University of Pennsylvania. He continued his study of *Paramecium* there, taught undergraduates, and mentored graduate and postdoctoral students for the next twenty years and rose through the ranks to professor.

With a National Science Foundation Post Doctoral Fellowship he went for a year's study of *Paramecium* with Professor Geoffrey Beale in his laboratory at the University of Edinburgh and with Dr. Dorothea Widmayer of Wellesley College, Massachusetts, who was also on sabbatical there. They focused on the isolation and characterization of the endosymbionts present in the different stocks of paramecia. Preer became proficient with the electron microscope, thanks to Dr. Artur Jurand at the university.

He returned to the United States in 1968 to accept a professorship in biology at Indiana University in Bloomington. In 1977 he was made distinguished professor of biology. He served as chairman of the first Department of Biology (the previous Departments of Zoology, Botany, and Microbiology were combined) from 1977 to 1979. His last sabbatical was spent with a Guggenheim Fellowship in Dr. Joseph Gall's laboratory at Yale, where he was exposed to research in the stimulating field of molecular biology. His research from then on centered on the molecular biology of *Paramecium*. He has approximately one hundred scientific publications and has mentored a handful of postdoctoral, thirty or more doctoral, and a few master's students, and has taught hundreds of college students.

His most recognized work was on "killers" in *Paramecium*. When killers were first discovered by Sonneborn, in the late 1930s, the trait was found to be caused by genetic determinants in the cytoplasm called kappa. Preer and his students were able to show that kappa was one of many symbiotic bacteria living within *Paramecium*. They also were the first to isolate the antigens of the serotype system of *Paramecium* and worked on a number of other traits.

At a Gordon Conference in Cold Spring Harbor in 1984, three different laboratories, one in Germany working on *Stylonichia*, one in France working on *Paramecium*, and Preer's laboratory in the United States, all reported that the universal genetic code of Watson and Crick was not quite universal as far as the stop colons were concerned. *Paramecium* uses as a stop only UGA of the three universal stop codons, UAA, UAG, and UGA. For many ciliates UGA is the only stop codon. In *Paramecium* UAA and UAG code for the amino acid glutamine. This was the first deviation from the universal genetic code known in the genome of a eukaryote at that time.

In 1967 Preer organized the first conference on *Paramecium* and *Tetrahymena* attended by scientists from Europe and the United States at Shelter Island, New York. He participated in international congresses of the Society of Protozoologists, and was president of the society in 1986. For many years he presented papers at meetings concerning the biology and molecular biology of ciliates that took place in Europe and Asia as well as at Gordon and FASEB conferences in the United States. He has been a member of the National Academy of Sciences since 1976. In 1993 he received the honorary doctorate in mathematics and natural sciences, University of Westfälische Wilhelms-Universität, in Münster, Germany.

1 Introduction

1.1 INTRODUCTION

Genetics is divided into classical Mendelian mechanisms and epigenetic mechanisms. Epigenetic phenomena include, by definition, all phenomena that cannot be accounted for by simple Mendelism. Most of them show cytoplasmic inheritance, but they turn out to be unrelated and operate by different mechanisms. Nowhere are there more numerous and more varied instances of epigenesis than in *Paramecium*. It is clear that *Paramecium* is a very useful organism for the study of epigenetic phenomena and may be expected to enlarge and enrich our knowledge of genetics proper in the future. Although crosses between strains with alternative characters provide the basis for both classical Mendelian genetics and epigenetics, numerous additional techniques are used for epigenetic analysis. In this book we will explore all these phenomena.

1.2 GENETICS

The principles of classical genetics were originally founded on the experiments of Mendel with the garden pea *Pisum sativum*. The results were first published in 1865, though they were not known to the scientific world for some time, until they were rediscovered in 1900 by G. Correns, H. deVries, and A. von Tschermak. Genetics was also based on the biological theories of A. Weismann,[1] who suggested that biological materials are separable into two parts: a "plasmon," or hereditary part, which could never be formed anew, but only by reduplication of something existing previously, and a "soma," or nonhereditary part. The word *genetics* was originally used by Bateson,[2] who defined it as the "study of the phenomena of heredity and variation."

Mendel found that there was no blending of the factors determining classical inheritance, as had been believed previously by Darwin and others. Mendel found that the factors that determine heredity segregated cleanly from each other at germ cell formation. These factors are now called genes. Early work by Strasburger, Hertwig, and Boveri in the 1880s led them to the conclusion that the chromosomes, which are found in the nucleus, have an important connection with heredity. The chromosome theory of heredity was introduced in the early 1900s, when it was shown in *Drosophila* (Sutton, Sturtevant, Morgan, Muller, and Bridges; see Carlson,[3] pp. 31–32) that the behavior of the chromosomes in meiosis and mitosis paralleled the behavior of genes.

Our whole outlook on genetics has been changed radically in recent years by the opening of the molecular era by Beadle and Tatum,[4] who studied the biochemical genetics of the fungus *Neurospora*, and numerous others working on bacteria. Avery, Macleod, and McCarty[5] showed that the chemical nature of the substance mediating the transfer of genetic characteristics in bacteria was DNA, rather than—as previously thought—a protein. Finally, Crick, Franklin, Wilkins, and Watson unraveled the chemical structure of DNA.[6] The structure of DNA provided a mechanism for its replication, and the triplet code explained the activity of DNA in specifying proteins. The Watson-Crick theory is now widely accepted and has resulted in a veritable explosion of biochemical and genetic experiments. The genetics of human beings has also been much studied recently in the Human Genome Project. The complete DNA sequence of many organisms is now known, including *Paramecium* and *Tetrahymena*. The range and variety of the biological material investigated for genetics has in consequence of all this work been enormous.

1.3 PARAMECIUM

This book is devoted to *Paramecium*—a unicellular ciliated protozoan, described by Haeckel.[7] It belongs to the phylum Protozoa and has a separate nucleus and cytoplasm, each surrounded by a membrane. It is therefore classified as a eukaryote.

Early workers in the seventeenth and eighteenth centuries, such as Leeuwenhoek, using primitive microscopes, first observed microorganisms like *Paramecium*.[8,9] Fundamental work on *Paramecium* in the nineteenth century was done above all by two outstanding European scientists, Maupas[10,11] in Algiers and Hertwig[12] in Munich, both of whom obtained important information about its sexual processes. In addition, in the nineteenth and early years of the twentieth century, many other biologists (e.g., Auerbach, Enriques, Chatton, Woodruff, Bütschli, Hartmann, Calkins, Erdmann, Dobell, Jollos, Jennings, and others too numerous to name here) devoted considerable time and attention to the protozoa. Their work was concerned with the significance of forms of life consisting of single cells, their life cycles, including especially the sexual stages, and the supposed immortality and rejuvenescence of such cells.

A fundamental aspect of genetics is the role of variation. Without variation there would be no genetics in any organism, for the genetic method itself is dependent upon the study of alternative characters. Some variations are due to genetic differences, but others are caused by past exposure to different environments. How do we distinguish between hereditary and environmental variations? We simply grow the organisms bearing the alternative traits in the same environment and see whether the traits are constant from one generation to the next before we pronounce the difference hereditary. In multicellular organisms we measure adult generations, each consisting of about eight or nine cellular generations in the worm *Coenorhabditis elegans*,[13] and thirty-six cellular generations in *Drosophila*.[14] On the other hand, in single-celled organisms we deal with cellular heredity, not with the genetics of whole organisms. For how many cell generations must a difference persist before we pronounce it hereditary? An answer to this question has been proposed by Preer.[15,16] He notes that one can calculate that if a cell the size of *Paramecium* consisted of all hydrogen molecules, it would have about 2^{60} molecules (the maximum number of molecules any cell could have), and all cell lines would therefore undergo a complete molecular turnover every sixty fissions. Therefore, he proposed that one take sixty fissions as the number of cellular generations that should be used in practical experiments to test whether a difference is hereditary or environmental. If this limit is also

Introduction

applied to the germline in multicellular organisms, one would have to estimate that it would take several generations of adult organisms before sixty cellular generations in the germline were obtained.

One reason that *Paramecium*, as contrasted with more familiar organisms—multicellular plants and animals—is desirable for studying genetics is that it is not in the mainstream of evolution of animals, plants, and humans. Ciliates are unique in the structure of both their nuclei and their outer surface, or cortex. They have two kinds of nuclei: very small micronuclei and the much larger macronuclei. The micronucleus acts as the generative nucleus, and the genes in the micronucleus are generally not expressed, while the macronucleus is the somatic nucleus and its genes are expressed, taking care of the normal day-to-day cellular functions. (The reason for different functions of the two kinds of nuclei is discussed in Chapter 3.) The micronuclei give rise to new macronuclei at every autogamy and conjugation, as described in Chapter 3; the micronucleus is 2n in chromosome number, while the macronucleus is estimated at about 860n. At the reproductive processes of conjugation and autogamy, the micronuclear chromosomes are broken into smaller sections and numerous sections (internal eliminated sequences [IESs]) are deleted. See Chapter 8. Thus, the chromosomal complement undergoes a wholesale reorganization when new macronuclei are formed from micronuclei during the sexual processes of conjugation or autogamy.

As stressed by Frankel,[17] ciliates are very peculiar organisms. Not only do they have two kinds of nuclei, but also the structural organization of *Paramecium* and other ciliates is perpetuated longitudinally as the ciliate cell grows. The arrangement of structures in the cortex is determined by copies of structures in front and behind them in a longitudinal fashion. These are only two of the many peculiar features of *Paramecium*, compared with the more familiar organisms used in the study of genetics, such as peas, *Drosophila*, *Escherichia coli*, and humans.

1.4 THE EARLY WORKERS

In 1929 Jennings wrote a lengthy review entitled *Genetics of the Protozoa*.[18] Since mating types were not known at that time, the analyses were based on cellular (uniparental not biparental) inheritance, and hence there could not be much genetics in that account as we understand it today. It was thought by some biologists then that the principles of classical genetics did not apply to the protozoa. Jennings himself[18] stated that the possibility of Mendelian inheritance in the protozoa was "largely speculative." Similar ideas were expressed by T. M. Sonneborn (Figure 1.1), who did most of the early work on the genetics of *Paramecium*, and who distanced himself from the conventional beliefs of geneticists in the first part of the twentieth century. In his autobiographical notes Sonneborn recalled some remarks made by his undergraduate advisor Andrews at Johns Hopkins University in 1922, that the current excitement about chromosomes and genes was a temporary fashion that would pass in time. (See Beale[19] and Sonneborn.[20,21]) Even a decade later Sonneborn and Lynch[22] wrote that "the present state of knowledge of the genetics of protozoa, unlike the genetics of higher organisms, does not justify the assumption of genes." On the other hand, it was Sonneborn himself who was soon to correct this error by

FIGURE 1.1 Tracy M. Sonneborn. Courtesy Indiana University Archives.

the discovery of mating types in what is now known as *P. primaurelia*. See Chapter 2. This was a most important finding, for it enabled one to do typical cross-breeding analyses in *Paramecium*. Soon after this he discovered the first gene in *Paramecium*. By cross-breeding it was determined that this gene behaved in typical Mendelian fashion. See Chapter 5.

The early concentration of *Paramecium* workers on non-Mendelian heredity was probably the result of the special interests and beliefs of a few people who worked with this organism, especially Sonneborn. Having previously worked on the genetics of flowering plants, one of the authors of this book (G. H. B.) was first attracted to microorganisms because they were expected to give quicker results than the plants or animals previously studied. The life cycle of most plants takes at least a year, and of *Drosophila* about 10 days, while the bacterium *E. coli* divides about every 60 minutes, and *Paramecium* about every 5 hours. When he asked the famous geneticist

Introduction 5

Barbara McClintock where he should go for a postdoctoral, she recommended Sonneborn as "the most original of American geneticists." As a graduate student the other author (J. R. P.), after a course in protozoology with Sonneborn (in which he covered only through the flat worms), elected to study *Paramecium* under Sonneborn. This decision was made although he had published on studies of insects (*Thysanoptera*) and had come to Indiana University and been accepted by A. C. Kinsey to work on gall wasps. Sonneborn was a charismatic scientist whose enthusiasm for *Paramecium* research was indeed contagious, and students from all over the world came to work with him at Indiana University.

Sonneborn throughout his life remained a dominant figure in the genetics of *Paramecium*. As already noted, his discovery of mating types led the way to a serious study of genetics in the protozoa. Although he was able to show numerous gene mutations that acted in a completely normal Mendelian fashion, he uncovered many examples of cytoplasmic phenomena, where the traits simply followed the cytoplasm, rather than segregating according to Mendel. These included mating types in several of the species in the *Paramecium aurelia* group, antigenic traits, cortical inheritance, symbiosis as exhibited in kappa particles and their relatives, and many other examples. At one point he tried to explain all these examples with the plasmagene hypothesis—the notion that copies of each gene are replicated in the cytoplasm. This will be discussed in Chapter 6. As a result of this early work, a substantial amount of information about *Paramecium* was obtained, some of which was very controversial and not acceptable to other geneticists; in spite of that, however, much interest was aroused. After all, Sonneborn, notwithstanding his unorthodox views, was at one time president of the Genetics Society of America (in 1949), and his opinions were considered seriously by most geneticists. To do his work, a collection of laboratory cultures of *Paramecium* from all over the world was established and classified, and extensive knowledge of how to handle this material was built up, mainly by workers in Sonneborn's own group at Indiana University. In 1953, one of us (G. H. B.) wrote a small book, entitled *Genetics of* Paramecium aurelia, about some of this early work.[23] That book is now out of print and out of date, and was largely concerned with the importance of genetics based on the cytoplasm, rather than on the nucleus, although Sonneborn himself thought that the book was oriented too much toward orthodox genetics based on the nucleus, in spite of my efforts at that time to draw attention to the cytoplasm. However, this hypothesis was abandoned when the symbiotic nature of one of his examples, kappa, was demonstrated. It soon became apparent that the cytoplasmic phenomena that he was observing were a heterogeneous group, consisting of symbiosis, cortical structures, competing pathways of protein synthesis (serotype determination), and other unexplained phenomena.

1.5 DISCARDED HYPOTHESES

An unusual feature of this book is the detailed accounts of experiments bearing on hypotheses that were at one time much discussed by *Paramecium* workers, but which have been subsequently rejected as being unsound, though they seemed justifiable to some of us at one time on the basis of earlier information. We refer here particularly to the plasmagene and metagon hypotheses. See Chapter 6. Such accounts are usually excluded from modern science books, and may seem to some readers to be only of historical interest. Therefore, some may want to skip these sections, though we consider them important. Some of our predecessors were very accomplished scientists who had good reasons for their erroneous beliefs, as they are now seen to be. Descriptions of some of their work will give younger readers an idea of how science has developed in the past, and may even arouse healthy suspicions that the permanence of some current biological dogmas is not absolute.

1.6 EPIGENETICS

The term *epigenetics* as used by biologists has two different meanings. According to Waddington,[24] it is defined as the genetics of development. According to Lederberg,[25] it is a term that includes all inherited changes not due to the sequence of bases in DNA. We define an epigenetic trait as one that does not follow Mendelian laws of inheritance, but shows an alternative pattern of inheritance, usually cytoplasmic. As expected, epigenetic mechanisms are very diverse. For example, the mechanism of one case, DNA methylation,[26] consists of the stable transmission of the methylated state versus the unmethylated state of bases in DNA from one cellular generation to the next. Second, in cortical inheritance in *Paramecium*, the mechanism is totally different and results from the fact that the cortical structures of the outer surface of *Paramecium* determine the number and position of newly forming cortical structures. See Chapter 10. A third example is the inheritance of antigenic expression in *Paramecium*, which seems to depend upon steady states of metabolism much like those originally suggested by Delbruck.[27] See Chapter 7. Epigenetic mechanisms (as defined by Lederberg) are especially common in embryogenesis in plant and animal cells, and may well be the basis for most of the differential changes seen in development. (Perhaps one day we will find that Waddington's definition and Lederberg's definition will turn out to be virtually identical!) Other examples are mitochondria and chloroplasts, which have their own DNA (see Chapter 6), are inherited by cytoplasmic inheritance, and are thought to have evolved from free-living organisms that began as symbionts in the cytoplasm. Although these structures are not inherited in a Mendelian fashion, but instead show cytoplasmic inheritance, they are not strictly epigenetic factors either, since their inheritance is ultimately dependent on the sequences of bases in their own DNA. Kappa particles of *Paramecium* also have their own DNA and are symbionts that cannot be cultured outside of the cytoplasm. They seem to occupy a niche between organelles like mitochondria or chloroplasts and free-living organisms like the green algae of *Paramecium bursaria* that can live either symbiotically or free.

1.7 CONCLUSIONS

Much work has been done on *Paramecium*, and a number of publications, partly or wholly devoted to that organism and to some other ciliates, have been published, among which may be mentioned books or articles by Preer,[16,28] Gall,[29] Meyer and Beisson,[30] Grell,[31] Van Wagtendonk,[32] Sonneborn,[21] Kaneshiro et al.,[33] Nanney,[34]

Wichterman,[8,9] Görtz,[35] Frankel,[17] Bleyman,[36] Grimes and Aufderheide,[37] Hausmann and Bradbury,[38] Kung et al.,[39] and others too numerous to mention.

There was a time recently when interest in *Paramecium* had declined, leading to a paucity of laboratories around the world studying it. See, for example, the article by Preer[40] entitled "Whatever Happened to *Paramecium* Genetics?" In fact, most modern textbooks on genetics or molecular biology written at that time did not even include the word *Paramecium* in their indexes, and biology students of today are therefore unlikely to be aware of its importance. *Paramecium* can now be cryopreserved in liquid nitrogen by a technique described by Simon and Schneller,[41] and a fairly comprehensive selection of the stocks of the organisms used by Sonneborn is maintained in the American Type Culture Collection at ATCC, 10801 University Boulevard, Manassas, Virginia 20110-2209. In any case, we agree that some of the hereditary features of *Paramecium* do not fit in well with conventional ideas of either classical or molecular genetics as now understood.

However, some very interesting new work on ciliate genetics is going on at the present time, and there now seems to be a revival of interest in the subject, involving especially the transformation of the organisms by injecting DNA into the macronucleus (where apparently any DNA can replicate), the interrelations between micro- and macronuclei, the "gene silencing" phenomenon, homology-dependent inheritance (described in Chapter 12), and the use of a method of gene cloning by functional complementation[42-44] (see Chapter 4).

Apart from *Paramecium*, a substantial amount of research today is in progress on some other ciliate protists, notably *Tetrahymena*. Indeed, more scientific work is probably being done at present on *Tetrahymena* than on *Paramecium*. In any case, it should be stressed that the present book is far from a complete account of research even on the one genus *Paramecium*, and we apologize to the *Paramecium* workers whose researches have been neglected here. We have made a drastic, and no doubt personally biased, selection of old and new work, perhaps unduly emphasizing aspects that may seem to have little relation to studies with the more familiar organisms used in most genetic research today.

In conclusion, we must point out that we certainly do not wish to deny the importance of much excellent recent work on conventional DNA genetic systems. There are plenty of examples of that kind of work from *Paramecium* itself, as will be shown in this book. We feel that it is important to describe other mechanisms that are particularly evident in *Paramecium*. A number of these will be described and will be referred to as instances of epigenetic inheritance. See Chapter 12. Some of these examples may well have their application to organisms other than *Paramecium*, and may be important in human genetics. In general, therefore, this book should be most interesting to readers who are prepared to entertain unorthodox views on genetics.

REFERENCES

1. Weismann, A. 1893. *The germ plasm: A theory of heredity.* English translation by Parker, W. N. and Rönfeldt, H. New York: Scribners, 1973.
2. Bateson, W. 1907. Inaugural address. In *Report of the Third International Conference 1906 on Hybridization and Plant Breeding*, 90. London: Royal Horticultural Society.
3. Carlson, E. A. 1981. *Genes, radiation and society.* Ithaca: Cornell University Press.

4. Beadle, G. W. and Tatum, E. L. 1941. Genetic control of biochemical reactions in *Neurospora*, *Proc. Natl. Acad. Sci. U.S.A.* 27:499.
5. Avery, O. T., Macleod, G. M., and McCarty, M. 1944. Studies on the chemical nature of the substance inducing transformation of pneumococcal types. *J. Exp. Med.* 79:137.
6. Watson, J. D., and Crick, F. H. C. 1953. A structure for deoxyribose nucleic acids. *Nature* 171:737.
7. Haeckel, E. 1866. *Generelle Morphologie der Organismen*. 2 vols., English trans. Berlin: G. Reimer.
8. Wichterman, R. 1953. *The biology of* Paramecium. 1st ed. New York: Blakiston.
9. Wichterman, R. 1985. *The biology of* Paramecium. 2nd ed. New York: Plenum Press.
10. Maupas, E. 1888. Recherches expérimentales sur la multiplication des infusoires ciliés. *Arch. Zool. Exp. Gén.* 2:165.
11. Maupas, E. 1989. La rajeunissement karyogamique chez les ciliés. *Arch. Zool. Exp. Gén.* 2:149.
12. Hertwig, R. 1889. Über die Conjugation der Infusorien. *Abh. Wiss. Bayer. Akad. Wiss.* 17:150.
13. Wilkins, A. S. 1992. *Genetic analysis of animal development*. 2nd ed. New York: Wiley-Liss.
14. Drost, J. B. and Lee, W. R. 1995. Biological basis of germline mutation: Comparisons of spontaneous germline mutation rates among *Drosophila*, mouse, and human. *Environ. Mol. Mut.* 25(Suppl. 26):48.
15. Preer Jr., J. R. 1958. Nuclear and cytoplasmic differentiation in the protozoa. In *Developmental cytology*, ed. D. Rudnick, 4. New York: Ronald Press.
16. Preer Jr., J. R. 1969. Genetics of protozoa. In *Research in protozoology*, Vol. 3, ed. T. T. Chen, 150. Oxford: Pergamon Press.
17. Frankel, J. 1989. *Pattern formation in ciliate studies and models*. New York: Oxford University Press.
18. Jennings, H. S. 1929. Genetics of the protozoa. *Bibl. Genet.* 5:108.
19. Beale, G. H. 1982. Tracy Morton Sonneborn. *Biog. Mems. Fell. R. Soc.* 28:537.
20. Sonneborn, T. M. 1947. Recent advances in the genetics of *Paramecium* and *Euplotes*. *Adv. Genet.* 1:263.
21. Sonneborn, T. M. 1975. Parmecium aurelia. In *Handbook of genetics*, Vol. 2, ed. R. C. King, chap. 20. New York: Plenum Press.
22. Sonneborn, T. M. and Lynch, R. S. 1932. Racial differences in the early physiological effects of conjugation in *P. aurelia*. *Biol. Bull.* 62:258.
23. Beale, G. H. 1953. *The genetics of* Paramecium aurelia. London: Cambridge University Press.
24. Waddington, C. H. 1942. The epigenotype. *Endeavor* 1:18.
25. Lederberg, J. 2001. The meaning of epigenetics. *The Scientist* 15:6.
26. Bird, A. 2002. DNA methylation patterns and epigenetic memory. *Genes and Development* 16:6.
27. Delbruck, M. 1949. Discussion in "Influence des gènes, des plasmagènes et du milieu dans le déterminisme des caractères antigéniques chez *Paramecium aurelia* (variété 4)." In *Unités biologiques douées de continuité génétique*, ed. A. Lwoff, 33. Vol. 8. Paris: CNRS.
28. Preer Jr., J. R. 2001. Genetics of *Paramecium*. In *Encylopedia of genetics*, ed. E. C. R. Reeve, 243. London: Fitzroy Dearborn.
29. Gall, J. G. 1986. *The molecular biology of ciliated protozoa*. New York: Academic Press.
30. Meyer, E. and Beisson, J. 2005. Epigenetics: *Paramecium* as a model system. *Med. Sci.* 21:377.
31. Grell, K. G. 1968. *Protozoologie*. 2nd ed. Berlin: Springer-Verlag.
32. Van Wagtendonk, W. J. 1974. *Paramecium: A current survey*. New York: Elsevier.

Introduction

33. Kaneshiro, E. S., et al. 1979. The fatty acid composition of *Paramecium aurelia* cells and cilia changes with culture age. *J. Protozool.* 26:147.
34. Nanney, D. 1980. *Experimental ciliatology: An introduction to genetic and developmental analysis of ciliates.* New York: Wiley.
35. Görtz, H. D. 1983. Endonuclear symbionts in ciliates. *Intern. Rev. Cytol.* 14:145.
36. Bleyman, L. 1996. Ciliate genetics. In *The ciliates: Cells as organisms*, ed. K. Hausmann and P. C. Bradbury, 292. Stuttgart: Fischer-Verlag.
37. Grimes, G. W. and Aufderheide, K. J. 1991. *Cellular aspects of pattern formation: The problem of assembly.* Basel: Karger.
38. Haussmann, K. and Bradbury, P. C. 1996. In *The ciliates: Cells as organisms*, ed. K. Hausmann and P. C. Bradbury, 1. Stuttgart: Fischer-Verlag.
39. Kung, C., et al. 2000. Recent advances in the molecular genetics of *Paramecium*. *J. Eukary. Microbiol.* 47:11.
40. Preer Jr., J. R. 1997. Whatever happened to *Paramecium* genetics? *Genetics* 145:217.
41. Simon, E. M. and Schneller, M. V. 1973. The preservation of ciliated protozoa at low temperature. *Cryobiology* 10:421.
42. Haynes, W. J., et al. 1996. Toward cloning genes by complementation in *Paramecium*. *J. Neurogenet.* 1:81.
43. Haynes, W. J., et al. 1998. The cloning by complementation of the *pawn-A* gene in *Paramecium*. *Genetics* 149:947.
44. Meyer, E. and Cohen, J. 1999. *Paramecium* molecular genetics: Functional complementation and homology-dependent gene inactivation. *Protist* 150:11.

2 Mating Types in *Paramecium*

This chapter is placed here at the beginning of the book because the concept of mating types is basic for the analysis of genetic and epigenetic aspects of *Paramecium*.

When the early work on *Paramecium* was being done during the nineteenth century, it was not possible to obtain sexual processes of *Paramecium* at will in the laboratory, although conjugation was observed from time to time by several observers. Thus, Maupas[1] remarked, very optimistically, "on peut s'en procurer des conjugaisons, autant que l'on désirera" [one is able to get conjugation whenever one wishes]. However, it was realized by early workers that without knowledge of how to get conjugation, genetic analysis by Mendelian methods was an impossibility with *Paramecium*. A major step forward was therefore taken by Sonneborn[2] when he finally made Maupas's observation a reality when he discovered at Johns Hopkins University how to obtain conjugation in *Paramecium* at will. It was considered by him (see Beale[3]) that the discovery of mating types in *Paramecium* was a major landmark in the progress of genetic research of ciliates. From that time, genetic analysis by Mendelian methods became possible, for one could then do what Sonneborn had previously considered his main task, namely, to find out to what extent a unicellular organism like *Paramecium* conformed with the rules of classical genetics, and to what extent other mechanisms operated. He had not forgotten what had been suggested to him when he was a student at Johns Hopkins University, that there might be alternative hereditary systems in lower organisms, different from those of genes and chromosomes. Now in 2007 it should be remembered that when such methods involved conjugation or autogamy in *Paramecium*, formation of new macronuclei from micronuclei takes place, and the macronuclei thus formed, which determine the phenotype, are not genic replicas of micronuclei (see Chapter 8). That would be a complication that inevitably arises when conjugation or autogamy is used as a mechanism for getting recombination in *Paramecium*, and possibly also other difficulties that were understandably not appreciated by Sonneborn in his early work. These will be discussed later.

Sonneborn's discovery of mating types in *Paramecium*[2,4] was based on his finding that some cultures often displayed many cells in the act of conjugation after what had originally been called endomixis, but was later shown to be autogamy (see Chapter 3). In the original work, single isolations of cells were made of what was then called *Paramecium aurelia* stock S (now called stock 19), which is now *P. primaurelia*, in group A of the *P. aurelia* group of species (see Chapter 5). A stock was defined as the descendants of a single isolation of a *Paramecium* brought in from nature. These isolations were allowed to produce separate clones. He then made mixtures of all possible combinations of the clones thus obtained. The result was that in many mixtures there was the immediate formation of many large clumps of cells and their subsequent separation into hundreds of individual conjugating pairs.

This discovery of how to get conjugation was very dramatic. Sonneborn, who made it while working in the middle of the night of March 1, 1937, was so excited at the time that he felt he simply must call someone else into his lab to see such an amazing event, and the only other person in the building of Johns Hopkins University at the time was, as he put it, "an old janitor." Sonneborn has described (see Beale[3]) how he dragged the janitor to the microscope and said, "Look at that mating reaction! Just look at it! Isn't it wonderful?" The janitor, according to Sonneborn's account, had probably never looked in a microscope before and may not have seen anything. But he said to Sonneborn, "Boy, it sure must be wonderful, because you're mighty excited." Sonneborn commented: "Indeed I was." (Actually, we now realize that it was fortunate that Sonneborn did this experiment with one of the group A species of *Paramecium* [see Chapter 5], otherwise he might not have been so lucky.)

Some combinations of clones, however, failed to produce any clumps or pairs at all. When the material examined contained a mixture of different stocks belonging to different species (at first called varieties), Sonneborn found that he could separate his ex-autogamous isolates into three species (species 1, 2, and 3), each containing two mating types. Thus, species 1 contained mating types I and II, species 2 contained mating types III and IV, and species 3 contained mating types V and VI. The reactions occurred as shown in Table 2.1: when mating type I was mixed with type II, or III with IV, or V with VI, and the environmental conditions were right, conjugation always occurred; but when two clones belonging to the same type (either all I, all II, all III, etc.) or to different species (I with III, I with IV, I with V, etc.) were

TABLE 2.1

Mating Reactions Among the Six Mating Types of Species 1, 2, and 3

		P. primaurelia		*P. biaurelia*		*P. triaurelia*	
	MT	I	II	III	IV	V	VI
P. primaurelia	I		+++				
	II	+++					
P. biaurelia	III				+++		
	IV			+++			
P. triaurelia	V						+++
	VI					+++	

From Sonneborn, "*Paramecium aurelia*: mating types and groups; lethal interactions, determination and inheritance," *Amer. Nat.*, vol. 73, p. 395, Table 4, 1939, with permission of the University of Chicago Press.

mixed, no clumps or pairs were seen. As a result of these experiments, Sonneborn was able to formulate the concepts of mating types and species (though the species were at first called varieties and the mating types denoted sexes[2]), but this nomenclature was short-lived, and the expression "mating types" was adopted instead of "sexes" and has been retained ever since. Some critics, for example, Darlington,[5] have argued in favor of other terms, such as "incompatibility groups." Jennings[6] also preferred the expression "incompatibility groups" to "mating types."

2.1 MATING TYPES AND THE SPECIES PROBLEM IN CILIATES

There first appeared to be three species (the mating types were denoted with Latin numerals, and species with Arabic numerals). In later work, more species were found among the stocks of *P. aurelia* collected from different parts of the world. By 1946 there were eight,[7] by 1957 sixteen (including some species of *P. multimicronucleatum*),[8] and by 1975 fourteen.[9] Later a fifteenth was discovered. Species 15 was later found to differ morphogically from the others and was removed from the *P. aurelia* group and called *P. sonneborni*,[10] leaving fourteen species. Probably more species and mating types of *P. aurelia* will be found in the future, though the most common ones are probably all known by now. Each of these species of *P. aurelia* contained no more than two mating types.

In some cases there are cross-reactions among the different mating types. See Table 2.2. The strongest cross-reactions occur between species 4 and 8. Thus, mating type VII of species 4 reacts strongly with mating type XVI of species 8, and mating type VIII of species 4 reacts strongly with mating type XV of species 8. However, while 90% of the VIII by XV crosses exchange nuclei, nuclei are exchanged in only 10% of the VII by XVI matings.[11] Such "interspecies" mating reactions,[8,9,12–14] for the most part, are much weaker than the homologous pairings. Moreover, either they die immediately or at the next autogamy they give rise to progeny that are not viable. The few remaining survivors tend to be like the parents that were originally mixed, so gene exchange between species is virtually nonexistent. Rarely, pairs of *P. aurelia* may sometimes be formed between two individuals classified as belonging to a single mating type. These are the "selfers." Selfers will be discussed in Chapter 5.

While species of *P. aurelia* contain only two mating types, other distinct species of *Paramecium* may contain more than two mating types, each mating with all the other types, but none with themselves. For example, *P. bursaria* was found to contain 2, 4, or 8 mating types in different subspecies,[6,15–17] as contrasted with only two in the *P. aurelia* complex.

Moreover, Sonneborn originally found that the species could not be clearly distinguished by any character other than mating type. It was known, however, that there were also, in addition to the mating type properties, a number of not precisely defined specific character differences, affecting cell size, fission rate, environmental conditions necessary for mating, lethal temperature, mode of inheritance of mating types, and so on, but none of these characters were sufficiently clear to be used for the unambiguous identification of individual species. Consequently, Sonneborn at first rejected the idea that his originally described "varieties" should be given the rank of "species."

TABLE 2.2
The Known Cross-Reactions Among the Mating Types

Gp	Sp	MT	A P. primaurelia O	A P. primaurelia E	A P. triaurelia O	A P. triaurelia E	B P. tetraurelia O	B P. tetraurelia E	A P. pentaurelia O	A P. pentaurelia E	B P. septaurelia O	B P. septaurelia E	B P. octaurelia O	B P. octaurelia E	B P. decaurelia O	B P. decaurelia E	B P. dodecaurelia O	B P. dodecaurelia E
A	P. prim-aurelia	O		+++										+-				
A	P. prim-aurelia	E			+ 2%						++ 2%		+-					
A	P. tri-aurelia	O				+++					+			++ 2%				
A	P. tri-aurelia	E																
B	P. tetra-aurelia	O						+++					+++ 90%	+++ 2%	+		+	
B	P. tetra-aurelia	E																
A	P. pent-aurelia	O							++ 40%	++ 40%								
A	P. pent-aurelia	E										+++						
B	P. sept-aurelia	O									+-							
B	P. sept-aurelia	E										+++						
B	P. oct-aurelia	O												+++	+		+	
B	P. oct-aurelia	E																
B	P. dec-aurelia	O														+++		
B	P. dec-aurelia	E																
B	P. dodec-aurelia	O																+++
B	P. dodec-aurelia	E																

Duplicate half of table omitted. Number of +s represent strength of mating reaction. % represents proportion of cells in a mixture forming true conjugants. Amount of death after autogamy in conjugants is not given, but is low

Modified from Sonneborn, T.M., *Paramecium aurelia*, in: *Handbook of Genetics*, 2nd ed., King, R.C., Ed., Plenum Press, chap. 20, p. 572, Fig. 1, 1975. With kind permission of Springer Science and Business Media.

Later, in an extensive paper on species problems in protozoa, Sonneborn[8] stated that, according to the then modern species concept,[18,19] a species was defined as a group of organisms within which crossing, with genic exchange, could occur. Obviously the species of *P. aurelia* are sexually isolated from each other. It should be pointed out that intraspecies crosses of different stocks (a stock being a group of organisms all derived from a single organism, collected at a given site in nature), though they produced F1s that were fully viable, frequently resulted in the later formation of F2 progeny containing a sometimes very high (even 90–99) percent of clones that were not viable. Such incompatibilities have been shown to be due probably, as will be shown later, to variations in chromosomal number between stocks (see Chapter 8).

In the same paper,[8] p. 303, Sonneborn wrote "The difficulties and hazards involved in identification clearly preclude assigning species names to the varieties." He also rejected serological characters and combinations of other traits. He concluded by stating that he would continue to use the term *species* in its universal pigeonhole sense and find a new term for the limited number of cases in which common gene pools were known. He suggested the word *syngen* (p. 303)[8] for a group of organisms within which gene exchange occurred. At that time (1957), and for some years thereafter, he and other *Paramecium* workers used the term *syngen* in place of *variety*.

But, after a further eighteen years, Sonneborn took up the question of varieties and syngens yet again, and as a result he changed the nomenclature of *P. aurelia* completely. He then wrote[13] as follows: "The situation has recently changed. Almost all the 'syngens' can now be identified without recourse to any standard living cultures. Thirteen of the fourteen 'syngens' can be routinely identified by the electrophoretic patterns of a small number of cytoplasmic or mitochondrial enzymes (Tait, 1970,[20] Allen and Gibson, 1971,[20a] Allen, Byrne, and Cronkite, 1971,[21] Allen, Farrow, and Golembiewski, 1973[22]). The time has come, therefore, as others, e.g., Hairston, 1958,[23] have long maintained—prematurely as it seemed to me, ... to recognize each 'syngen' as a 'species,' which implies conferring on it a species name."

However, he was not quite ready to take that step even then, and in his comprehensive review[13] he merely changed *syngens* to *species*, respectively, without giving them Latin names. But, later in the same year he wrote a definitive paper[9] entitled "The *Paramecium aurelia* Complex of Fourteen Sibling Species." In that paper he declared that the time had come when identification had become practicable and Linnaean names should be given to the species. He declared that the designation of *P. aurelia* was a *nomen dubium* and stated (p. 158): "So I propose to assign to each species a name that includes both '*aurelia*' and its former numerical designation, as follows: *P. primaurelia, P. biaurelia, P. triaurelia, P. tetraurelia, P. pentaurelia, P. sexaurelia, P. septaurelia, P. octaurelia, P. novaurelia, P. decaurelia, P. undecaurelia, P. dodecaurelia, P. tridecaurelia* and *P. quadecaurelia*, all new species." They correspond to the previously designated variety, syngen, or species 1–14, respectively. In this book we shall refer to the species as *P. primaurelia, P. biaurelia, P. triaurelia*, and so on, or for the sake of brevity, species 1, 2, 3, and so forth.

Sonneborn's reputation was so great among the *Paramecium* workers that most of them immediately accepted this somewhat unconventional nomenclature. The numbering of the mating types with Latin letters as I, II, III, IV, and so on, was for a time retained, though later (see Sonneborn[13]) it was realized that since there were two mating types in each species, one could be called O (odd) and the other E (even). The designations of O and E were made such that the E in all stocks was homologous to all the other Es, and the O in all stocks was homologous to all the other Os. Although these assignments were made primarily on the basis of cross-reactions, Sonneborn used other criteria as well. Thus, mating type I could be called O, mating type II could be called E, mating type III could be called O, mating type IV could be called E, and so on.

The groups of species *P. primaurelia*, *P. biaurelia*, and so on, were denoted by Sonneborn as the "*P. aurelia* species complex." This classification into species complexes, or sibling species, though unusual, is a remarkable achievement in the taxonomy of *Paramecium*. Such an arrangement has not been made with any other organism, as it would involve a vast and impractical amount of crossing work.

Some information on the geographical distribution of the different species complexes of *P. aurelia* about the world has been published. Some data of this kind are given by Sonneborn,[8,13] Wichterman,[24] Przybos,[25,26] and others. The species *P. primaurelia* and *P. biaurelia* are found all over the world, but some other species are apparently distributed only locally; for example, *P. novaurelia* was found originally only in Europe, and *P. sexaurelia* was originally found by Sonneborn only in tropical regions, though Przybos[25,26] later reported that *P. sexaurelia* was also found in Spain, and had previously been reported from other localities in 1975. *P. quadecaurelia* has been found only in Australia.[9] Individual regions, apart from the United States, have been sampled only very incompletely. Some data published by Polish and Russian authors indicate that certain species are more widely distributed than has been previously believed.

It should be added that the means whereby *Paramecium* moves from one place to another are unknown. The organism cannot withstand drought and has never been known to form cysts, yet is found widely over the world. *Paramecium* may be carried from one expanse of water to another by birds or other animals, though this has never been proved conclusively. An alternative hypothesis is that *Paramecium* is such an old species that it was around before the continents separated from Pangea, and it has really not moved around very much, but only as the continents have moved.

To this account of the distinctions observed between the different species within the *P. aurelia* complex by mating, we may add that there is now a much quicker, molecular method for identifying the various sibling species (*P. primaurelia*, *P. biaurelia*, etc.), as compared with the previous laborious method of testing them for their ability to form conjugants with known representatives of standard mating types and observing whether true conjugation takes place. This new method is by making random amplified polymorphic DNA–polymerase chain reaction (RAPD-PCR) fingerprints. See Stoeck et al.[27] for using the method on *Paramecium*. The method requires, as testers of individual species, samples of as few as three individual *Paramecium* cells (or sometimes only a single one). Using such a method, a single pair of primers was shown to be sufficient to identify nine sibling species of the *P. aurelia* complex (of the 14 now known), and further work will probably show that the remaining sibling species will also be distinguishable by this method.

RAPD-DNA markers have also been used by Tsukii[28] for subdividing the morphospecies (species that are morphologically distinct) of *P. caudatum*, though the method there was found not to be so satisfactory as with the *P. aurelia* complex, because the known different mating type groups or syngens of *P. caudatum* do not comprise such a neat system as the mating types of the *P. aurelia* complex. *P. caudatum* and other species of *Paramecium* have been left till now with the syngen system of notation.

The taxonomic system finally adopted by Sonneborn can undoubtedly be justified for the *P. aurelia* complex, and it fits that group very well, but if we move from *P. aurelia* into other ciliate species complexes such as *P. caudatum*, apparently insoluble nomenclatural problems are evident. The species *P. bursaria* and other ciliate genera, such as *Tetrahymena*, *Euplotes*, and so on, also reveal serious problems in their systems of mating type and species diversification, by comparison with the *P. aurelia* complex. (See, for example, discussion of the situation in *P. caudatum* by Hiwatashi[29] and Tsuki.[30]) Sonneborn left to others the decision whether to revise the nomenclature of ciliate taxa other than *P. aurelia*. So far, none have been altered, except *Tetrahymena*. In that case, however, even more radical changes than for *P. aurelia* have been made by Nanney and McCoy.[31] The numerical system of syngens in *Tetrahymena* was abandoned altogether by Nanney and McCoy, and each species was described by a conventional Linnaean binomial designation, such as *Tetrahymena thermophila*, *T. pigmentosa*, and so forth.

The consequences of Sonneborn's discovery of mating types in *Paramecium* were profound. As he pointed out, the system of mating types and species, and the understanding of how to obtain conjugation, enabled him to start an analysis by Mendelian methods of the differences between genotypes of particular species of *P. aurelia*. This had been his wish for many years previously (see Beale[3]), but did not prevent him (Sonneborn) from studying variations of *Paramecium* that represented epigenetic difference. See Chapter 12. In Chapter 5, the genetic, or epigenetic, relations of mating types in the *P. aurelia* complex are discussed. These experiments have revealed new problems, many still unresolved. It should be emphasized, however, that all subsequent work on the genetics of *Paramecium* depends heavily on the analysis of mating types.

REFERENCES

1. Maupas, E. 1889. La rajeunissement karyogamique chez les ciliés. *Arch. Zool. Exp. Gén.* 2:149.
2. Sonneborn, T. M. 1937. Sex, sex inheritance and sex determination in *P. aurelia*. *Proc. Natl. Acad. Sci. U.S.A.* 23:378.
3. Beale, G. H. 1982. Tracy Morton Sonneborn. *Biog. Mems. Fell. R. Soc. Lond.* 28:537.
4. Sonneborn, T. M. 1939. *Paramecium aurelia*: Mating types and groups; lethal interactions, determination and inheritance. *Am. Nat.* 73:390.
5. Darlington, C. D. 1955. Review of Beale, 1955. *Heredity* 9:284.
6. Jennings, H. S. 1939. *Paramecium bursaria*: Mating types and groups, mating behavior, self sterility; their development and inheritance. *Am. Nat.* 73:414.
7. Sonneborn, T. M. and Dippell, R. 1946. Mating reactions and conjugation between varieties of *P. aurelia* in relation to conceptions of mating type and variety. *Phys. Zool.* 19:1.

8. Sonneborn, T. M. 1957. Breeding systems, reproductive methods, and species problems in protozoa. In *The species problem*, ed. E. Mayr, 155. Washington, D.C.: AAAS.
9. Sonneborn, T. M. 1975. The *Paramecium aurelia* complex of fourteen sibling species. *Trans. Am. Microsc. Soc.* 94:155.
10. Aufderheide, K. J., Daggett, P.-M., and Nerad, T. A. 1983. *Paramecium sonneborni* n. sp., a new member of the *Paramecium aurelia* species complex. *J. Protozool.* 30:128.
11. Haggard, B. W. 1974. Interspecies crosses in *Paramecium aurelia* (syngen 4 by syngen 8). *J. Protozool.* 21:159.
12. Sonneborn, T. M. 1947. Recent advances in the genetics of *Paramecium* and *Euplotes*. *Adv. Genet.* 1:264.
13. Sonneborn, T. M. 1975. *Paramecium aurelia*. In *Handbook of genetics*, ed. R. C. King, chap. 20. Vol. 2. New York: Plenum Press.
14. Beale, G. H. 1954. *The genetics of* Paramecium aurelia. London: Cambridge University Press.
15. Siegel, R. W. and Larison, L. L. 1960. The genetic control of mating types in *Paramecium bursaria*. *Proc. Natl. Acad. Sci. U.S.A.* 46:344.
16. Chen, T. T. 1963. New mating types of *Paramecium bursaria*. *J. Protozool.* 10(Suppl.):22.
17. Bomford, R. 1966. The syngens of *Paramecium bursaria*. New mating types and intersyngenic mating reactions. *J. Protozool.* 13:497.
18. Mayr, E. 1940. Speciation phenomena in birds. *Am. Nat.* 74:249.
19. Mayr, E., ed. 1957. *The species problem*, 155. Washington, D.C.: AAAS.
20. Tait, A. 1970. Enzyme variation between syngens in *Paramecium*. *Biochem. Genet.* 4:461.
20a. Allen, S. L. and Gibson, I. 1971. Intersyngenic variations in the esterases of axenic stocks of *Paramecium aurelia*. *Biochem. Genet.* 5:161.
21. Allen, S. L., Byrne, B. C., and Cronkite, D. L. 1971. Intersyngenic variations in the esterases of bacterized *Paramecium aurelia*. *Biochem. Genet.* 15:135.
22. Allen, S. L., Farrow, S. W., and Golembiewski, P. A. 1973. Esterase variations between the 14 syngens of *Paramecium aurelia* under axenic growth. *Genetics* 73:561.
23. Hairston, N. G. 1958. Observations on the ecology of *Paramecium* with comments on the species problem. *Evolution* 12:440.
24. Wichterman, R. 1985. *The biology of* Paramecium, chap. 1. 2nd ed. New York: Plenum Press.
25. Przybos, E. 1990. Interspecies differentiation of *Paramecium sexaurelia* (*Ciliophora*). *Arch. Protistenkd.* 138:123.
26. Przybos, E. 1996. Interspecies differentiation of *Paramecium triaurelia*. Strains from Romania. *Folia biologica* (Krakow) 44:1.
27. Stoeck, T., Pzrybos, E., and Schmidt, H. J. 1998. A combination of genetics with inter- and intra-strain crosses and RAPD-fingerprints reveals different population structures within the *Paramecium aurelia* species complex. *Eur. J. Protistol.* 34:348.
28. Tsukii, Y. 1996. Genetic diversity among natural stocks of *Paramecium caudatum* revealed by RAPD markers. *Eur. J. Protistol.* 32(Suppl. I):165.
29. Hiwatashi, K. 1968. Determination and inheritance of mating type in *Paramecium caudatum*. *Genetics* 58:373.
30. Tsukii, Y. 1988. Mating type inheritance. In *Paramecium*, ed. H.-D. Görtz, 59. Berlin: Springer-Verlag.
31. Nanney, D. L. and McCoy, J. W. 1976. Characterization of the species of the *Tetrahymena pyriformis* complex. *Trans. Am. Microsc. Soc.* 95:664.

3 General Description of the Protista and of *Paramecium*

3.1 INTRODUCTION

It is the aim of this chapter to describe briefly the nature of the material that will be dealt with in the rest of this book.

The term *protozoa* (which today includes only the single-celled animals) was coined by Goldfuss in 1820.[1] Goldfuss included some metazoa as well as single-celled animals in his protozoa. Haeckel[2] classified all organisms into three groups: plants, animals, and the single-celled organisms, or Protista. Later Haeckel[3] revised his system and made it explicit that the single-celled protozoa were included in the Protista. Protists have often been considered by biologists to be the first living things existing on this planet, from which protozoa and multicellular animals and plants have evolved. In former times, university courses in biology started with brief descriptions of the protozoans, amoeba and *Paramecium*, with the implication that they were simple, unicellular organisms and were at the first stage of evolution toward higher multicellular animals and plants. Now we recognize DNA as the indispensable component of nuclei, and it is held to be the foundation of genetics. Most textbooks on biology now begin with the structure and function of DNA.

We open this chapter with some remarks about the possible origin and evolution of Protista. As stated above, it was initially believed by many biologists that Protista like those now existing were the first forms of life on the earth. However, though widely accepted by many, this view was strongly opposed by Dobell.[4] Dobell prefaced his article with the provocative statement that the present paper is "largely analytic and destructive: but it is so of necessity, for it is useless to build on a rotten foundation." Dobell expressed the view that it was incorrect to describe the Protista as primitive organisms. In Dobell's view, nobody who had worked with protozoa for any length of time believed that they were very simple. Enough was known even in 1911 for it to be realized that the life cycles of the amoebas were complex. There were dozens of species of amoeba, each with a different life cycle. Was any one of them really simple? Dobell thought that nobody who tried to work out one life cycle of one amoeba would answer that question in the affirmative. "The truth, however," he stated (p. 307),[4] "is that the Protista are very small—but they are not simple," and we might add that this applies not only to amoeba but also to *Paramecium*, the main organism that will be dealt with in this book. We now recognize that the simple protist from which the higher forms were thought to issue is a hypothetical simple ancestor, and the complex organisms we see today were derived from that ancestor by a long period of evolution leading more often than not to great complexity. Indeed, the protists of today are undoubtedly complex, as readers may well agree after reading some of the later chapters of this book, or considering some of the properties of *Paramecium* as described in the books of Wichterman[5,6] and others.

Living organisms are now classified into two groups: eukaryotes, those with nuclei surrounded by nuclear membranes in the cytoplasm, and prokaryotes, those without. This division is almost universally made by biologists today. Are ciliates such as *Paramecium* eukaryotes? Chatton[7] was probably the first to consider whether protists were "procaryotes" or "eucaryotes" (spelled with a *c*). Chatton placed the ciliates in the class of eucaryotes in that paper. Responsibility for the terms *eukaryote* and *prokaryote* (spelled with a *k*) is now usually given to Stanier and Van Niel.[8] Today the term *eukaryote* is usually taken to mean an organism in which the genetic material is contained in one or more structures called nuclei and separated from the cytoplasm by membranes. *Paramecium* certainly qualifies as a eukaryote, for both the micronucleus and the macronucleus are surrounded by nuclear envelopes. Eukaryotes are thought to have arisen from the simpler prokaryotes.

A recent issue of the *Journal of Eukaryotic Microbiology* is devoted entirely to an article attributed to twenty-eight coauthors[9] entitled "The New Higher Level Classification of Eukaroytes with Emphasis on the Taxonomy of Protists." It is based on data from modern morphological studies, biochemical pathways, and molecular phylogenetics, and the authors modestly assert that they "believe the classification will have some stability in the near term." Six major clusters of eukaryotes, each made up of smaller related groups, are found in Table 1 of the paper on p. 45. The multicellular animals are found in the Opisthokonta, and the multicellular plants are in the Archaeplastida. The remaining four groups (with some typical forms included in parenthesis) are made up of Protista. They include the Amoebozoa (including the amoebas), the Rhizaria (including the foraminifera and radiolaria), the Chromalveolata (including the ciliates and *Paramecium*), and the Excavata (including the flagellates). These six groups are all thought to have arisen from prokaryotic ancestors, the eubacteria or true bacteria and the archaea or archibacteria.

It will be assumed in this book that organisms of the *P. aurelia* species group consist of cells, though objection has been made to use of the word *cell* for *Paramecium* by some biologists, including Dobell. In using this word here, we do not wish to imply that the "cells" of *Paramecium* are exactly like those of the differentiated cells of multicellular plants or animals. In a previous book on *Paramecium*,[10] individual paramecia were called animals. In the past there had been much argument about the use of the word *cell* for individual protists. See Dobell,[4,11] Lwoff,[12] and Corliss.[13–15] It should be noted that due to their complicated structure and the variety of functions they must muster into a single individual, they are really not like multicellular organisms, whose cells tend to be much more specialized. Nevertheless, we need a short word for referring to individual paramecia, and for lack of any alternative, in this book we will call them cells rather than, as formerly, animals.

In a book like this one, concerned with genetics and epigenetics of *Paramecium*, it should be noted that the hereditary variations of single-celled organisms are equivalent to two different types of variation in multicellular organisms. The genetics of multicellular organisms is based on the traits of adults from one generation to the next.

Such studies have produced the classical picture of genetics. The differences between the various cell types found in a single multicellular organism, however, are usually considered more appropriate for embryological rather than for genetic study. Since these differences occur in the cells of embryos, it is impossible to tell whether they are hereditary or simply environmentally produced, for they cannot be cultured outside the embryo in a constant environment like protozoa. In protozoa, where cells can be cultured independently, such distinctions can be readily made, and show how useful single-celled organisms are for studying variations, both genetic and epigenetic.

Finally we must define a stock of *Paramecium* (see also Chapter 2). Numerous collections of *Paramecium* have been made over the years and have been maintained in the laboratory. When individuals are taken into the laboratory from nature and an isolate is made, its progeny are called a stock. Numerous stocks of *Paramecium* are now available for biological studies and are maintained in various laboratories around the world and in the American Type Culture Collection at ATCC, 10801 University Boulevard, Manassas, Virginia 20110-2209.

3.2 MORPHOLOGY

A full description of the morphology of *Paramecium* is given by Wichterman.[6] It should, however, be mentioned that in comparison with other microorganisms such as bacteria, yeasts, microscopic algae, and so on, the cells of protists belonging to the *P. aurelia* group are unusually large (see Figure 3.1). They are about 150 μm long and 50 μm broad, but vary considerably in size, according to food supply, stage in the life cycle, and the particular stock under study. They are free-living organisms, not parasites, and live in freshwater ponds, streams, and rivers throughout the world. Many collections have been made over the years. *Paramecium* is covered in thousands of cilia, which provide locomotion. *Paramecium* has a macronucleus, two or more micronuclei, and two contractile vacuoles that occur on the aboral (dorsal) surface. These vacuoles are located about the first and last quarters of the body and empty their fluid contents to the exterior by means of excretory pores, which are fixed structures in the cortex. On the side of the cell denoted "ventral," there is a groove, or gullet, into which food (bacteria, yeasts, algae, etc.) is conveyed by special cilia. At the bottom of the gullet, the food is concentrated into food vacuoles that are continuously budded off and enter the cytoplasm. In living cells these food vacuoles continuously circulate around in the cytoplasm. The undigested residue of food is later discharged into the exterior medium through an anal pore (denoted the cytopyge or cytoproct). The cytopyge is ventral and subterminal, and rather close to the posterior end of the body.

3.2.1 THE CORTEX

The cortex is a highly complex structure and comprises the outer surface of *Paramecium*. It will be described briefly here. For details, the reader is referred to Beisson and Sonneborn,[16] Jurand and Selman,[17] Grimes and Aufderheide,[18] Frankel,[19] Iftode et al.,[20] Hausmann and Bradbury,[21] Sonneborn,[22] Adoutte,[23] and Chapter 10 of this book. The cortex of *Paramecium* consists of cortical units (see Figure 3.2) repeated several thousand times. These units are repeated longitudinally along the surface of

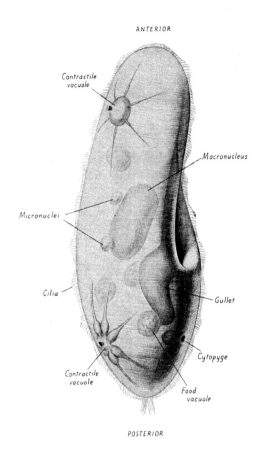

FIGURE 3.1 *Paramecium* of the *P. aurelia* complex. Figure shows various structures, including two micronuclei and a single macronucleus. Reproduced with permission of Palgrave Macmillan from Jurand, A., and Selman, G. G., *The Anatomy of* Paramecium aurelia (London: Macmillan, 1969), 85, plate 1.

the *Paramecium*, and the longitudinal row of units is called a kinety (Figure 3.2). Figure 3.3 shows a silver-stained *Paramecium*. The term *kinetosome* (see Lwoff[12]) is used here instead of the frequently used "basal body" or "basal granule." A single cortical unit consists of a cilium (see Figures 3.4 and 3.5) attached to a kinetosome (often a second kinetosome is present, with or without a cilium, depending on its state of development); systems of fibers, especially kinetodesmal fibers, which are long, tapered, overlapping fibers attached to the base of the kinetosome that join others and run longitudinally toward the front of the cell (see Figure 3.6); and finally a parasomal sac (Figure 3.7). A trichocyst is usually inserted between each of the units that make up a kinety. Newly forming kinetosomes are especially interesting, for they develop in a precise location in relation to other structures within the asymmetric unit, at right angles to the surface, and gradually move 90 degrees to an upright position and eventually develop cilia (see Figure 3.8).

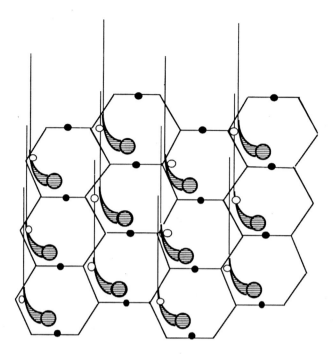

FIGURE 3.2 A simplified representation of the location of structures in the cortex of *Paramecium*. Twelve hexagonal ciliary units are shown. Each has a large cross-hatched circle that represents a kinetosome. A kinetodesmal fiber originates from its base and runs to the left and forward for several units until it tapers off and terminates. The resulting bundle of fibers separates the units on their sides. Each long column of units (shown here are three units) is called a kinety. Kineties often run the length of the whole *Paramecium* and consist of many units. A trichocyst insertion site (small solid circle) is on the border of each unit. The small open circle just above and to the left of each kinetosome is the parasomal sac.

The kinetosomes at one time were thought to contain DNA[24] and to be self-reproducing genetic structures, or what was called by Lwoff[12] "unités douées de continuité génétique." However, it is now clear that they do not contain DNA,[22,25] and their supposed self-reproducing properties are part of the larger aspect of the positional role of structures in the development of the cortex.

In times of stress, such as exposure of *Paramecium* to toxic chemicals or hostile predators, the trichocysts are discharged into the external medium as long striated filaments with pointed tips (Figures 3.9 and 3.10). Ever since the time of Maupas,[26] when he remarked, "Le rôle des trichocystes offensifs est si évident," it has been assumed that trichocysts performed some protective defensive function against predators of *Paramecium*, but such a function was actually unproven until recently, when it was shown experimentally that the presence of extrusible trichocysts protects *Paramecium* from being devoured by the ciliates *Dileptus* and *Monodidinium*.[27-29] *Paramecium*, of course, lives in a freshwater environment containing many predators, including some voracious ciliates such as *Didinium*, *Monodidinium*, and *Dileptus*,

FIGURE 3.3 *Paramecium* silver stained. Ventral surface is seen here showing the pattern of kineties around the mouth, indicated by the arrow. Silver stains the bases of the cilia (kinetosomes) and the kinetodesmal fibers. The posterior of the cell is at the top of the figure. Reproduced with permission of Palgrave Macmillan from Jurand, A., and Selman, G. G., *The Anatomy of* Paramecium aurelia (London: Macmillan, 1969), 87, plate 2.

for which *Paramecium* is the usual, and indeed indispensable, food. The genetics of trichocysts is described in Chapter 10.

Near to the mouth there is a structure that appears during conjugation and autogamy, called the paroral cone, named by Diller.[30] This is a protuberance that plays an important role in the passage of the gamete nuclei from one conjugant to the other during conjugation.

3.2.2   THE CYTOPLASM

The cytoplasm contains numerous mitochondria that have genetic importance and will be considered in Chapter 6. There are also large numbers of bacterial endosymbionts (kappa and other particles) in the cytoplasm of many stocks of *Paramecium* collected from nature. These symbionts have genetic importance and will be described in detail in Chapter 6. *P. bursaria* has symbiotic algae. Many ribosomes can also be seen with the electron microscope and are considered in Chapter 9.

3.2.3   THE NUCLEI

The nuclei of *Paramecium*, like those of other Protista, are among its most remarkable features. Members of the *P. aurelia* complex of species contain one large

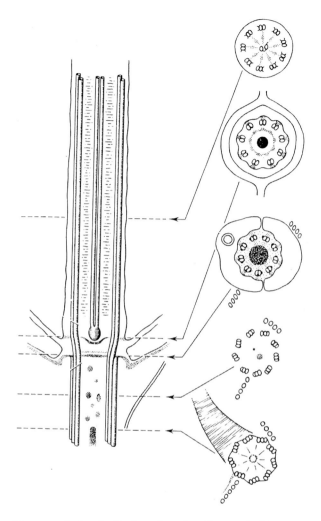

FIGURE 3.4 Diagram of a longitudinal section of a cilium and a kinetosome. The appearance of the cilium is shown in cross section at various levels. At the very bottom of the figure the kinetodesmal fiber is seen coming off the kinetosome and running forward. All sections except the topmost one are through the kinetosome. Reproduced with permission of Palgrave Macmillan from Jurand, A., and Selman, G. G., *The Anatomy of Paramecium aurelia* (London: Macmillan, 1969), 19, diagram 4.

macronucleus and (usually) two minute vesicular micronuclei. These nuclei comprise probably the most important genetic features of *Paramecium*. Some early workers, for example, Hickson[31] (p. 372), preferred the term *meganucleus* to the more usual *macronucleus*, pointing out that the Greek prefix *mega* means "big" while *macro* means "long." Use of the term *meganucleus* was also continued for a while by some early protozoologists, such as Dobell,[11] as well as by some early textbook writers, for example, Mackinnon and Hawes,[32] but today most protistologists now prefer the term *macronucleus*.

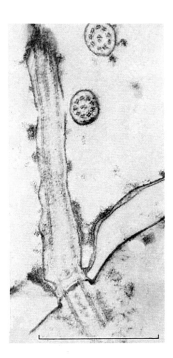

FIGURE 3.5 Longitudinal section through a cilium, and two cilia in cross section. One sees the familiar 9 + 2 pattern of microtubules in the cilium. The microtubules run the whole length of the cilium. Although this figure was originally prepared by Mott to illustrate antibody on the ciliary surface, it shows the structure of the cilium nicely. The bottom bar is 1 micron long. With permission from Mott, M. R., *J. Gen. Microbiol.*, 41, 260ff., 1965. All illustrations are on plates and follow plate 3, Figure 5.

The macronucleus of members of the *P. aurelia* complex of species is about 50–60 μm long and 20–30 μm broad, while the two micronuclei are each about 3 μm in diameter. The micronuclei are often located in a groove in the macronucleus.

Rarely, samples of the *P. aurelia* complex are found that contain more than two micronuclei, or sometimes by contrast none at all,[33,34] but usually two can be seen. However, *Paramecium* cannot live indefinitely without a micronucleus. Strains that have been artificially enucleated invariably die out after a number of fissions. Of the many hundreds of stocks established in the laboratory over the years, all have micronuclei. Their presence and vesicular structure are specific features of the *P. aurelia* complex.

The micronuclei of members of the *P. aurelia* complex contain, in different stocks, from 30 to more than 60 pairs of extremely small chromosomes, when observed at metaphase of the first meiosis.[35] See Figure 3.11. The actual number is characteristic of the stock.

As we will see in Chapter 8, micronuclei are inactive, while macronuclei contain the active genes of the cell. Moreover, macronuclei have their DNA cut into smaller pieces when they are formed from the micronuclei at autogamy and conjugation. During this process of macronuclear formation, amplification and deletion of sections of DNA also occur. The DNA increases from the micronuclear number (2n)

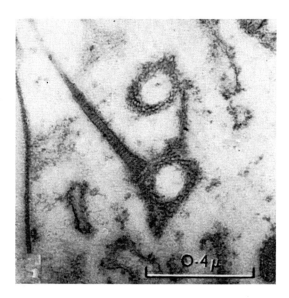

FIGURE 3.6 Two kinetosomes, showing a kinetodesmal fiber coming off the base of the older, bottom kinetosome and running forward. Reproduced with permission of Palgrave Macmillan from Jurand, A., and Selman, G. G., *The Anatomy of* Paramecium aurelia (London: Macmillan, 1969), 105, plate 11, Figure 5.

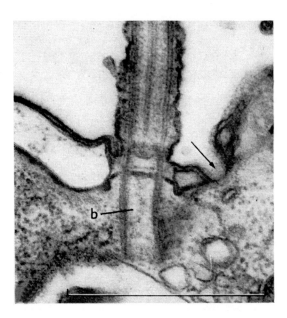

FIGURE 3.7 Longitudinal section showing parasomal sac. Figure shows a section of the surface of *Paramecium* showing the basal body (b) and the parasomal sac (unlabeled arrow). Line at bottom of figure is 1 micron. Reproduced with permission of Palgrave Macmillan from Jurand, A., and Selman, G. G., *The Anatomy of* Paramecium aurelia (London: Macmillan, 1969), 97, plate 7, Figure 3.

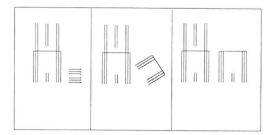

FIGURE 3.8 Diagram of a side view of three stages in the development of a kinetosome. In the left panel the new kinetosome is just forming at right angles to the old. In the middle panel the new kinetosome is moving into an upright position. In the last panel it has gained its proper position in front of the old kinetosome. Eventually the membranes and fibers of a new cilium will form.

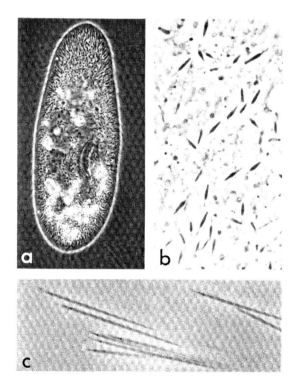

FIGURE 3.9 Trichocysts. (a) Hundreds of undischarged trichocysts can be seen in the *Paramecium*, mostly around the edge of the specimen. (b) Undischarged trichocysts can be seen in squash of cells; magnification about ten times that of (a). (c) Extruded trichocysts in medium surrounding cell—magnification like that of (b). With kind permission of Springer Science and Business Media, from Adoutte, A., in *Paramecium*, edited by H.-D. Görtz (New York: Springer-Verlag, 1988), 327, Figures 1a–c.

General Description of the Protista and of *Paramecium*

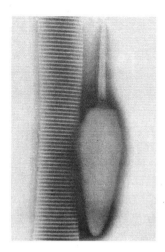

FIGURE 3.10 Negatively stained trichocysts. Electron micrograph of discharged and nondischarged trichocysts negatively stained with phosphotungstic acid. With kind permission of Springer Science and Business Media, from Adoutte, A., in *Paramecium*, edited by H.-D. Görtz (New York: Springer-Verlag, 1988), 329, Figure 2.

to the macronuclear number (860n). Phenomena that result in changes in DNA when the micronuclei make the macronuclei are described in Chapters 5, 8, and 12.

3.3 REPRODUCTION

3.3.1 ASEXUAL FISSION

When in presence of food *Paramecium* grows by a process whereby there is a gradual increase in size of the whole cell, until a constriction appears around the central

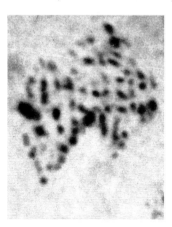

FIGURE 3.11 Chromosomes on the first metaphase plate. Chromosomes of stock 144, of *P. primaurelia*. Light microscope photograph. Reproduced with permission of Palgrave Macmillan from Jurand, A., and Selman, G. G., *The Anatomy of* Paramecium aurelia (London: Macmillan, 1969), 171, plate 44, Figure 7.

region, and the two halves then separate as two eventually identical daughter individuals. In general, under the most favorable conditions, that is, when there is a surplus of food and a temperature of about 27°C, asexual fission of the *P. aurelia* complex occurs at intervals of about 5 hours, but the optimum temperatures for growth and fission vary considerably according to the stock and species, and the stage in the clonal cycle.

During fission the two diploid micronuclei in each specimen of the *P. aurelia* complex undergo mitosis, though the cytological details of this process usually cannot be made out clearly in the *P. aurelia* complex owing to the small size of the chromosomes. However, it may be assumed that each micronucleus in a *Paramecium* divides mitotically once, and the products constitute the micronuclei of the two daughter cells. Mitosis, which takes place at regular intervals in the micronuclei, provided there is excess food, ensures that exact replicas of each gene and each chromosome are formed, and each of the two chromosomes of a pair is represented once in each of the daughter nuclei.

On the other hand, the macronucleus forms two new macronuclei by a process described as amitosis. This amitosis exhibited by the macronucleus is probably the best case of true amitosis seen in any organism.

As the macronucleus does not divide mitotically, there is no assurance that each macronuclear "chromosome" is represented the same number of times in each daughter nucleus. If there is not some invisible mitotic-like mechanism operating during macronuclear division, then the genic balance in the daughter macronuclei might be expected to change as the number of fissions increase. Since the phenotype is primarily determined by the macronucleus and not by the micronucleus, numerous changes in characters might be expected. How ciliates maintain genic balance without mitosis is one of the major mysteries of ciliate biology. See Chapter 8 for a discussion of these matters.

3.3.2 Conjugation

Conjugation in *Paramecium* has been studied for many years and cytological details of the process were long ago described by Maupas[34] and Hertwig.[36] But the genetic consequences of the process could not be understood at that time, because Mendel's laws were not known to the scientific public till 1900 at the earliest, nor was the chromosome theory of heredity, which developed after 1914, mainly from work of the Morgan school with *Drosophila*.

Nevertheless, Hertwig thought that conjugation should be considered mainly as a mechanism for bringing about sexual recombination of hereditary elements, while Maupas emphasized its role in rejuvenescence. Now we know that both were probably right, as will be discussed below, though there was considerable argument between these two workers at the time when they were both working on the problem.

Conjugation in *Paramecium* involves the pairing of two cells that may or may not be visually identical, but must always belong to different mating types, as discussed in Chapter 2. The process is shown diagrammatically in Figure 3.12 and described in detail in Chapter 4, Section 4.4. The conjugating cells pass through a complicated series of sexual processes lasting some hours.[37] After conjugation is

General Description of the Protista and of *Paramecium*

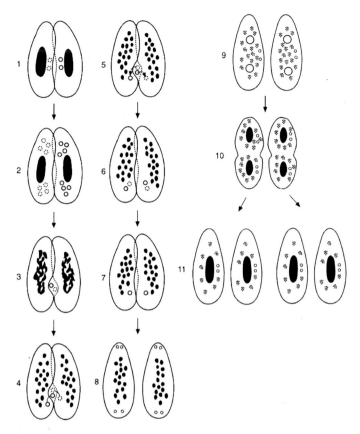

FIGURE 3.12 Nuclear changes at conjugation in *Paramecium*. (1) Conjugants start with a macronucleus and two micronuclei. (2) Each of the two diploid micronuclei in each conjugant undergoes two meiotic divisions to produce eight haploid nuclei (four are shown at the front and four at the back of each cell). (3) The macronucleus in each cell forms a skein-like structure. One of these micronuclei from each cell enters the paroral come, and those remaining are lost. (4) The macronucleus continues its degenerative process (which it will continue to the end of this diagram), breaking into 20–40 fragments. The two haploid micronuclei, one from each cell, undergo another mitotic division, producing two nuclei in each of the two paroral cones. (5) One of the two nuclei in each cell becomes the stationary nucleus and the other becomes the migratory nucleus. (6) The nuclei are now exchanged to give two haploid micronuclei in each cell. (7) Fusion of the two nuclei occurs to produce a single diploid nucleus in each cell. (8) The diploid nucleus in each cell now divides twice to produce four diploid nuclei. The cells separate. (9) Two of the nuclei in each cell now begin the process of forming two new macronuclear anlagen, while the other two remain micronuclei (small open circles). (10) Each of the two cells begins cell division. (11) The two micronuclei undergo mitosis to maintain their number at two per cell, while the two macronuclear anlagen are segregated one to each cell. The four cells produced in this way from separately formed macronuclei, along with their asexually produced offspring, are called caryonides—four caryonides from each conjugating pair. Thus, the normal vegetative condition of the cells is restored. At subsequent cell divisions the micronuclei undergo mitosis, and the macronuclei simply pinch in two amitotically. The macronuclear fragments are reduced in number by half at each cell division and are eventually lost. Reprinted with permission of Cambridge University Press from Beale, G. H., *The Genetics of* Paramecium aurelia (Cambridge: Cambridge University Press, 1954), 27, Figure 1.

over, the two participants swim apart and resume their separate lives. Genically, as will be described, they will have become identical. Conjugation is really a remarkable process occurring in ciliates but not in other groups of organisms.

When cells of different mating types but of the same species are mixed together, and cultural conditions are suitable, there is an immediate clumping of the cells into large groups. Sonneborn[37] has described some of the necessary environmental conditions. They involve a partial but not too severe deprival of food, certain limits of temperature, sometimes certain hours of light and darkness, and a certain "age" of the culture (i.e., time since the previous conjugation or other sexual processes). Some of these conditions are characteristic for particular species.[38]

After the mass clumping has been in progress for about 1 hour (at 27°C in *P. tetraurelia*), the groups of ciliates separate out into pairs, each pair containing one cell of each mating type. The individual cells in the pairs during conjugation are at first attached at their anterior ends (hold-fast union) and later, also in the region of the mouths (paroral cone union), into pairs, each pair containing one cell of each mating type. When they become attached in the region of the mouths, conjugants cannot be separated by force.[39] After the clumps containing many cells have broken up, the pairs swim about together until shortly after nuclei are exchanged.

After the conjugants have been paired for some time, the mouths become permanently closed, and eventually the gullets are reduced to rudiments. Consequently, the organisms do not feed during conjugation and, partly for this reason, become reduced in size as the process goes on. Meanwhile, the interior contents undergo a radical reorganization. According to Hertwig,[36] new mouths and gullets are formed during conjugation, by "budding" from rudiments of old ones (as also occurs during asexual, binary fission, but with the difference that at conjugation the old mouths and gullets are eventually lost completely). The contractile vacuoles, however, continue to function normally throughout conjugation.[5]

The process of conjugation normally lasts in *P. tetraurelia* for 5–6 hours at 27–29°C. Exceptionally, however, the cells may remain attached in pairs for longer periods at the paroral cone region, but not at the anterior hold-fast region. Indeed, sometimes (rarely) the cytoplasmic connection between the conjugants at the paroral cone region never breaks, and permanent doublets are formed.

The micronuclear changes that take place during and after conjugation of *Paramecium* are extremely complicated. They are not entirely clear from cytological observation alone, and it is necessary to make use of combined cytological and genetic data to complete the picture. In addition, inferences may be drawn from known cytological processes occurring during conjugation of ciliates other than *Paramecium*, since the basic cytological processes are similar in many ciliate genera.

The most obviously visible cytological change occurring during conjugation is the disintegration of the old macronucleus. During the second half of the period during which the conjugants are joined together, the macronuclear chromatin stretches out into a long skein, which then breaks into many more or less rounded fragments, the number of which may be as many as fifty-seven, according to Maupas, but in *P. tetraurelia* is more often about forty (see Figure 3.12, stage 4). These fragments of the old macronucleus are then distributed, without multiplying, to the exconjugants. The variability in duration of persistence of macronuclear fragments in the exconjugants is probably connected with differences in nutritional conditions during conjugation. When there is excess food, fragments of the old, disintegrating macronucleus pass into one or the other of the exconjugants and decrease in number per cell by half at each cell division. According to Preer (unpublished), all the fragments may be accounted for after ten fissions (1,024 cells), after which they become too rare to find. If, on the other hand, there is insufficient food, the fragments disintegrate and are lost much earlier.

At prophase of the first meiotic division the micronucleus elongates and becomes crescent shaped. It can be up to 20 microns in length. From studies with ciliates other than *Paramecium* having larger chromosomes, and from the general course of events, it is reasonable to infer that the first two micronuclear divisions constitute the prophase of meiosis, though most of the cytological details cannot be seen clearly in the *P. aurelia* complex.

After these two meiotic divisions of each of the two micronuclei, eight, presumably haploid, nuclei are formed in each conjugant. It is usually considered that seven of these eight nuclei disappear, though in some accounts it is stated that several of the eight may persist for a while. One remaining haploid nucleus in each conjugant then moves to the paroral cone.

The haploid nucleus in the paroral cone passes through a mitotic division, and as a result, two genically identical haploid gamete nuclei are formed in the paroral cone of each conjugant. One of the nuclei, the "male" in each conjugant, now crosses into the paroral cone of the opposite cell (see Figure 3.12, stage 5). The two nuclei fuse to form a diploid nucleus in each cell. Proof that the two haploid nuclei that participate in fertilization are derived from the two different members of a pair of conjugants was obtained by Sonneborn by experiments with genetic markers, using the first genes that he identified in *P. primaurelia*, the mating type alleles, *Mt* and *mt*, and the allelic genes determining presence or absence of the kappa particles, *K* and *k*, in *P. tetraurelia*.[40–42] Later work (by Beale[10]) has provided abundant confirmatory evidence for this exchange of nuclei during conjugation.

After fertilization, the diploid syncaryon in each of the two exconjugants divides twice mitotically, and separation of the two pairing cells usually occurs at some time between these two divisions. After the second division, each of the exconjugants contains four nuclei, two situated at the anterior and two at the posterior end of the cell (see Figure 3.12, stage 8). Two of these nuclei, according to Maupas,[34] probably the two anterior ones, transform into two new micronuclei. The other two increase enormously in volume and become very large, almost spherical, faintly staining bodies of diameter about 10 µm (see Figure 3.12, stage 9). Although the DNA is replicated during this period, the process is not completed until the cells have divided. They are the characteristic macronuclear anlagen (from the German for rudiments, singular is anlage), which eventually develop into normal, deeply staining macronuclei. One might suppose that these two anlagen would be identical since they arise from the fusion nucleus. However, it has been found that the clones, or caryonides, derived from each are often different. When they are, we speak of caryonidal inheritance. The transformation of micronuclei into macronuclei is indeed a remarkable phenomenon.

Sometimes, exconjugants develop more than two macronuclear anlagen. Sonneborn[40] reported that as many as ten anlagen may sometimes occur in certain lines. By a distribution of the extra macronuclei to the daughter cells without macronuclear division, the number is eventually restored to one per cell. It should be pointed out that the micronuclear genotype of all the progeny derived from a single pair is identical.

There remains the question of whether any significant exchange of cytoplasm takes place during conjugation. As mentioned above, the conjugants sometimes remain joined together by a strand of cytoplasm after completion of the normal process of conjugation. Sonneborn found that the longer the members of a pair stayed together at the paroral cone region after separation of the conjugants at the anterior hold-fast regions, the more cytoplasm was exchanged. He did this by measuring the amount of kappa that was exchanged. A rough measure of the amount of kappa exchanged was taken to be the time it took lines to regain strong killing after conjugation of a killer with a sensitive cell.[41,43,44]

The conditions in nature that bring about persistence of a connecting cytoplasmic strand between conjugating cells, and consequent exchange of cytoplasm between conjugants, are unknown. It is, however, possible to induce such persistence by a number of artificial treatments, such as exposure of conjugants to weak homologous antiserum. For details see Sonneborn.[45] Furthermore, exchange of cytoplasm is more likely to occur during conjugation involving certain stocks rather than others. For example, Sonneborn[43] found that exchange of cytoplasm often occurred when *P. tetraurelia* stock 51 conjugated with stock 29 or 47, but not when stock 51 conjugated with stock 32.

3.3.3 Autogamy

In addition to conjugation, all members of the *P. aurelia* complex undergo another important process called autogamy. The cytological events of autogamy are just like those of conjugation, except that they occur in single cells. Each autogamous cell produces two anlagen and two caryonides. Since there is no partner in autogamy, the two haploid gametic nuclei produced simply fuse together to form new diploid fertilized micronuclei, one in each cell and homozygous at all loci. These micronuclei go on to form new macronuclei. The old macronuclei are lost. Formerly, autogamy was called parthenogenesis by Hertwig[46] or endomixis by Woodruff and Erdmann.[47] We now know that endomixis—a process in which new macronuclei are formed from micronuclei without meiosis—does not occur, as shown by Diller.[30] Diller carried out a careful cytological study and left little doubt about the matter. This conclusion was confirmed genetically by Sonneborn,[42,48] using the fact that while endomixis should not lead to any change of genotype, autogamy should lead to segregation and homozygosity.

Table 3.1[49] shows some results obtained by passing the double heterozygotes g^{60}/g^{90} d^{60}/d^{90} through autogamy. Clearly there is segregation of alleles and independent assortment of the two pairs of alleles, indicating the occurrence of meiosis. Furthermore, all of the cells obtained are homozygous for both pairs of alleles, in accordance with expectation. It is clear that cells that have passed through autogamy become homozygous for all their genes in a single step.

TABLE 3.1
Genetic Effect of Autogamy in Animals Heterozygous for Two Independent Pairs of Genes ($g^{60}g^{90}$ and $d^{60}d^{90}$)

Genotypes of progeny	Number of exautogamous clones
$g^{60}g^{60}\ d^{60}d^{60}$	17
$g^{60}g^{60}\ d^{90}d^{90}$	18
$g^{90}g^{90}\ d^{60}d^{60}$	19
$g^{90}g^{90}\ d^{90}d^{90}$	20
Other (heterozygotes)	0
Dead	118
Total	192

Modified from Beale, G., *Genetics*, vol. 37, p. 67, Table 5, 1952. With permission.

It is now known that autogamy can be induced in all stocks of the *P. aurelia* complex by growing appropriate strains (that is, strains that are at a certain clonal cycle stage in the life cycle process, whether homozygous or heterozygous) with abundant food for a certain period, and then depriving them of food. If the period of growth with excess food has been sufficiently long, and they are at the right stage in the clonal cycle (Section 3.4), nearly 100% of the cells will pass through autogamy when deprived of food. The mean period of time between one autogamy and the next, or between conjugation and autogamy, is characteristic for the species and stock.[50] For example, in most stocks of the species *P. primaurelia*, the interautogamous period is about 10–14 days, or fifty fissions under conditions of rapid growth followed by starvation at 27°C, while in *P. tetraurelia* it is much shorter, 3–4 days, or fifteen fissions.

The above statements about when autogamy occurs apply to all species of the *P. aurelia* complex. Autogamy does not occur naturally in *P. caudatum*, though it can be induced there by special treatments.[51] In several other species autogamy has been recorded, though it is not thought to occur in *P. bursaria* (p. 281).[6]

3.3.4 MACRONUCLEAR REGENERATION

As described above, in the *P. aurelia* complex both conjugation and autogamy result in the loss of the previously existing macronuclei, and the differentiation of new macronuclei from micronuclei. The new macronuclei may be genically different from the macronuclei of the previous generation. However, Sonneborn[42] showed that in the *P. aurelia* group of species, formation of new macronuclei could also occur in the absence of nuclear fusion and without the involvement of micronuclei. This can happen if the new macronuclear anlagen fail to develop into new macronuclei. If this happens, the old macronuclear fragments now take their place and tend to enlarge. When they have been diluted down to about one per cell, they now conveniently begin to divide again. Sonneborn called this process macronuclear regeneration (MR). Macronuclear regeneration has also been reported to occur in *P. caudatum*.[52] In *P. tetraurelia*, Sonneborn[42] found that exposure of the cells to a temperature of 38°C at the stage when anlagen are just beginning to form, during conjugation or autogamy, will inhibit normal formation of the macronuclear anlagen, thereby inducing macronuclear regeneration. The descendants of *Paramecium* that have undergone macronuclear regeneration have phenotypes exactly like those of the cells from which the macronuclear fragments have been derived. According to Sonneborn and Schneller (p. 36),[53] macronuclear regeneration can be induced in stocks of *P. tetraurelia* especially well when the mutant allele *am* (amacronucleate) is present.

By ingenious genetic manipulations Sonneborn[54] was able to exploit the occurrence of macronuclear regeneration to construct heterokaryons, that is, cells containing macronuclei of one genotype and micronuclei of another, and thereby demonstrate that the phenotype of *Paramecium* is controlled by its macronucleus, rather than by its micronuclei.[54] This experiment, which will be considered in more detail in Chapter 8, threw light on many important problems of *Paramecium* genetics.

3.3.5 CYTOGAMY

Another process called cytogamy was named by Wichterman,[55] working with *P. caudatum*. However, it also occurs in the *P. aurelia* complex. Conjugation fails, for after cells come together and conjugation begins, micronuclei are not exchanged. Each partner undergoes a separate autogamy. When studying conjugation, it is essential to have marker genes present to check that exchange of nuclei has actually occurred.

3.3.6 HEMIXIS

There remains a set of phenomena in which macronuclei occasionally extrude chromatin or break into two or more pieces. These poorly understood phenomena have been called hemixis[30] and seem to be abnormal occurrences that have little or no genetic effects.

3.4 THE LIFE CYCLE

The life cycle in ciliates was recognized very early in the history of the work on *Paramecium*. It has been studied by numerous workers over the years. See Sonneborn,[56,57] Beisson and Capdeville,[58] Aufderheide,[59] Gilley and Blackburn,[60] Takagi,[61] and many others. Members of the *P. aurelia* complex, during long continued periods of growth, pass through a succession of stages that are concerned with their ability to undergo various reproductive events, such as fission, conjugation, and autogamy.

After conjugation cells typically enter a stage of **immaturity**, when fission rate is high, but cells in many strains are unable to conjugate. Sonneborn states that immature periods in the *P. aurelia* complex of species typically last for twenty-five to thirty-five cell divisions, but in other species such as *P. bursaria*, immaturity may last six hundred or more fissions. In *P. tetraurelia* and *P. octaurelia* there are no immature periods following conjugation.

Maturity follows and cells conjugate well and continue to multiply at maximum fission rate. During maturity cells will undergo autogamy if starved. Early in the cycle the frequency of starved cells undergoing autogamy is low, but it increases with time, and once it starts it moves rather quickly from 0% to 100%. At this time it may even occur in constantly overfed daily isolation lines. Later, however, the percent induced decreases. The boundaries of maturity are defined by the relative timing of the two reponses to depletion of the food supply—conjugation or autogamy. During maturity the cells respond to a reduction in nutritive conditions by becoming mating reactive. If they lack an appropriate partner and thus do not mate, they soon cease to be capable of mating and may proceed to undergo autogamy when starved. Autogamy cannot be induced during the early stages of the maturity, but becomes possible at a later stage. As the period of maturity continues, the time intervening between mating reactivity and autogamy during food depletion becomes progressively less. Then comes a time when the stimulus of starvation induces autogamy before the cells have become capable of mating, and they do not become capable of mating until autogamy is completed. The onset of this response to food depletion defines the end of maturity. The duration of maturity varies among different stocks and species, from very few cell generations, for clones arising at conjugation, to seventy-five or more generations.[62]

Old age follows maturity. During old age many abnormalities are observed. Fission rate slowly drops and mortality at conjugation and autogamy increases. The percentage of cells starting and completing autogamy may also drop, and the death of cells going into autogamy may increase.[56] These changes are accompanied by an increase in the frequency of abnormalities in prefission morphogenesis, in growth and division of parts of the cell, and in micronuclear mitoses.[63,64] The number of micronuclei per cell becomes more variable, with an increasing frequency of cells with no micronuclei, as well as some with three or four micronuclei.[63,65] Great variation appears among ex-autogamous clones from senescent parents, fewer and fewer becoming "rejuvenated" when they undergo autogamy.[66,67] The frequency of production of nonviable progeny at autogamy rises from practically 0% to 100%. Thus, senescence is accompanied by increasing defects in the macronuclei, micronuclei, and cytoplasm. Finally, death eventually comes in all lines.

Sonneborn showed that in the *P. aurelia* complex autogamy could replace conjugation and prevent old age and death. In fact, autogamy seems to have the same rejuvenating effect as conjugation, except that after autogamy there is no immature period. This may be related to the fact that when autogamy is induced in the laboratory, it usually occurs late in the life cycle, while conjugation is usually induced earlier.[68] Immaturity is set to a new life cycle start by conjugation but not by autogamy.

Under standard conditions of daily reisolation at 27°C, with excess of bacterized organic medium, which suppresses autogamy, Sonneborn[56] reported that the maximum number of fissions between fertilization and death was approximately 350 in *P. primaurelia*, 300 in *P. biaurelia*, and 200 in *P. tetraurelia*.

Many early workers were intrigued by the problem of whether *Paramecium* can live forever without conjugation, or whether conjugation is necessary to rejuvenate them. In seeking to answer this problem, Woodruff grew *Paramecium* in isolation for many years without any conjugation occurring. He used strain 18 in the Sonneborn collection for this work. This finding has gained credibility since the discovery of mating types, for stock 18 is a member of the *P. biaurelia* species; it is mating type E, and its production of mating type O necessary for mating, is extremely rare. However, the stock does undergo autogamy, and Sonneborn showed that autogamy can replace conjugation in the life cycle.

No stable wild-type strains of the *P. aurelia* group lacking micronuclei have ever been found. However, strains lacking micronuclei often occur in the laboratory, or can be created by micromanipulation, but none ever survive for long periods. At this point we must conclude that *Paramecium* indeed cannot live indefinitely without replacing the macronucleus at either conjugation or autogamy. This is certainly not true for other ciliates such as *Tetrahymena*. Amicronucleate strains are well known in that genus and get along indefinitely. Preer and Preer[69] attempted to grow *Tetrahymena* under close surveillance in the laboratory, checking them at every fission in order to measure the rate of death and see if nonviable lines were produced in sufficient quantity to account for the elimination of imbalances in chromosome number. Using random computer models, they concluded that the rate of elimination of nonviable lines by death was much too low to account for these imbalances. They suggested that either macronuclear genetic units are distributed in a mitotic-like way or there must be a self-correcting mechanism, such that macronuclear units are able to sense their concentration and correct imbalances.

In old age autogamy in *Paramecium* is inducible in up to 100% of the cells within a day of the cells reaching the stationary phase, induced by starvation. Later, however, autogamy becomes more and more difficult to induce. Capacity to mate may appear sporadically, especially when autogamy is harder to induce. Generation time gradually increases from about 5 to about 24 hours, and then is followed by death.

The nature of the primary cause for these senescent progressions has been much discussed, and it has been suggested that old age results from changes in chromosomal imbalance or deterioration in the macronucleus, or alternatively, some effect in the cytoplasm. Kimball[70] first proposed that deterioration in aging is based on increasing chromosomal imbalance resulting from random distribution of chromosomes to daughter nuclei during amitotic divisions of the macronucleus. Crosses of old cells with young cells have not provided a clear answer, but Sonneborn[57] concluded that the primary senescent changes were in the cytoplasm, not in the macronuclei. Nevertheless, further study of this matter by Aufderheide,[59] involving transfers of macronuclei and cytoplasm by injection, led him to conclude that the basic cause of senescence was due, as originally thought, to macronuclear disorganizations of the chromosome set up during long-continued reproduction of the macronuclear chromosomes. It is probable that this matter requires further investigation, but the present view is probably in favor of macronuclear irregularities as the most likely cause of increasing disturbances in aged clones. That seems reasonable in view of the uncontrolled replication and distribution of the macronuclear chromosomes. Restitution to normality is brought about by conjugation or autogamy, when new macronuclei formed from micronuclei start a new cell cycle.

REFERENCES

1. Goldfuss, G. A. 1820. *Handbuch der Zoology*, section 1, page 3. Nürnberg: Johann Leonard Schrag.
2. Haeckel, E. 1866. *Generelle Morphologie der Organismen*, section 1, page 2. English trans. Berlin: G. Reimer.
3. Haeckel, E. 1905. *The wonders of life*. New York: Harper.
4. Dobell, C. C. 1911. The principles of protistology. *Arch. Protistenk.* 23:269.
5. Wichterman, R. 1953. *The biology of* Paramecium. 1st ed. New York: Blakiston.
6. Wichterman, R. 1985. *The biology of* Paramecium. 2nd ed. New York: Plenum Press.
7. Chatton, E. 1925. *Pansporella perplexa*, amoebien à spores, protegées des daphnies. Réflexions sur la biologie et la phylogénie des protozoaires. *Ann. Sci. Nat. Zool.* 8:5.
8. Stanier, R. Y. and Van Niel, C. B. 1941. The main outlines of bacterial classification. *J. Bacteriol.* 42:437.
9. Adl, S. M., et al. 2005. The new higher level classification of eukaryotes with emphasis on the taxonomy of protists. *J. Eukaryot. Microbiol.* 52:399.
10. Beale, G. H. 1954. *The genetics of* Paramecium aurelia. Cambridge: Cambridge University Press.
11. Dobell, C. C. 1914. A commentary on the genetics of the ciliate protozoa. *J. Genet.* 4:131.
12. Lwoff, A. 1948. *Les organites doués de continuité génétique chez les protistes*. Paris: Editions du CNRS.
13. Corliss, J. 1957. Concerning the "cellularity" or acellularity of the protozoa. *Science* 125:988.
14. Corliss, J. 1989. The protozoan and the cell: A brief twentieth century overview. *J. Hist. Biol.* 22:307.
15. Corliss, J. 1999. Annotational excerpts from Clifford Dobell's 88-year-old insightful classic paper, "The principles of Protistology." *Protist* 150:85.
16. Beisson, J. and Sonneborn, T. M. 1965. Cytoplasmic inheritance of the organization of the cell cortex in *Paramecium aurelia. Proc. Natl. Acad. Sci. U.S.A.* 53:275.
17. Jurand, A. and Selman, G. G. 1969. *The anatomy of* Paramecium aurelia. London: Macmillan.
18. Grimes, G. W. and Aufderheide, K. J. 1991. *Cellular aspects of pattern formation: The problem of assembly*. Basel: Karger.
19. Frankel, J. 1989. *Pattern formation. Ciliate studies and models*. New York: Oxford University Press.
20. Iftode, F., et al. 1989. Development of surface pattern during division in *Paramecium*. 1. Mapping of duplication and reorganization of cortical cytoskeletal structures in the wild type. *Development* 105:191.
21. Hausmann, K. and Bradbury, P. C., eds. 1996. *The ciliates: Cells as organisms*. Stuttgart: Fisher Verlag.
22. Sonneborn, T. M. 1970. Determination, development and inheritance of the structures of the cell cortex. *Symp. Int. Soc. Cell Biol.* 9:5.
23. Adoutte, A. 1988. Exocytosis, biogenesis, transport and secretion of trichocysts. In *Paramecium*, ed. H.-D. Görtz, 325. New York: Springer-Verlag.
24. Smith-Sonneborn, J. and Plaut, W. 1967. Evidence for the presence of DNA in the pellicle of *Paramecium. J. Cell Sci.* 2:213.
25. Nanney, D. L. 1980. *Experimental ciliatology*, 172. New York: John Wiley.
26. Maupas, E. 1883. Contribution à l'étude morphologique et anatomique des infusoires ciliés. *Arch. Zool. Exp. Gén.* 2:427.
27. Harumoto, T. and Miyake, A. 1991. Defensive function of trichocysts in *Paramecium. J. Exp. Zool.* 260:84.

28. Harumoto, T. 1994. The role of trichocyst discharge and backward swimming in escaping behavior of *Paramecium* from *Dileptus margaritifer*. *J. Eukaryot. Microbiol.* 41:560.
29. Miyake, A. and Harumoto, T. 1996. Defensive function of trichocysts in *Paramecium* against the predatory ciliate *Monodinium balbiani*. *Eur. J. Protistol.* 32:128.
30. Diller, W. F. 1936. Nuclear reorganization processes in *Paramecium aurelia*, with descriptions of autogamy and "hemixis." *J. Morph.* 59:11.
31. Hickson, S. J. 1903. Part 1, Introduction and protozoa, 2nd fascicle by Farmer et al. In *A treatise on zoology*, ed. E. R. Lancaster. London: Black.
32. Mackinnon, D. L. and Hawes, R. S. J. 1961. *An introduction to the study of Protozoa.* Oxford: Oxford University Press.
33. Geckler, R. P. and Kimball, R. F. 1953. Effect of x-rays on micronuclear number in *Paramecium aurelia*. *Science* 117:8.
34. Maupas, E. 1889. La rajeunissement karyogamique chez les ciliés. *Arch. Zool. Exp. Gén.* 2:149.
35. Sonneborn, T. M. 1977. Genetics of cellular differentiation; stable nuclear differentiation in eukaryotic unicells. *Ann. Rev. Genet.* 11:349.
36. Hertwig, R. 1889. Über die conjugation der Infusorien. *Abh. Bayer. Akad. Wiss.* 17:150.
37. Sonneborn, T. M. 1950. Methods in the general biology and genetics of *Paramecium aurelia*. *J. Exp. Zool.* 113:87.
38. Sonneborn, T. M. and Dippell, R. 1946. Mating reactions and conjugation between varieties of *P. aurelia* in relation to conceptions of mating type and variety. *Phys. Zool.* 19:1.
39. Metz, C. B. 1948. The nature and mode of action of the mating type substances. *Am. Nat.* 82:85.
40. Sonneborn, T. M. 1939. *Paramecium aurelia*: Mating types and groups; lethal interactions; determination and inheritance. *Am. Nat.* 73:390.
41. Sonneborn, T. M. 1943. Gene and cytoplasm. I. The determination and inheritance of the killer character in variety 4 of *P. aurelia*. II. The bearing of determination and inheritance of characters in *P. aurelia* on problems of cytoplasmic inheritance, Pneumococcus transformations, mutations and development. *Proc. Natl. Acad. Sci. U.S.A.* 29:329.
42. Sonneborn, T. M. 1947. Recent advances in the genetics of *Paramecium* and *Euplotes*. *Adv. Genet.* 1:264.
43. Sonneborn, T. M. 1944. Exchange of cytoplasm at conjugation in *Paramecium aurelia*, variety 4. *Anat. Rec.* 89:577.
44. Sonneborn, T. M. 1945. The dependence of the physical action of a gene on a primer and the relation of primer to gene. *Am. Nat.* 79:318.
45. Sonneborn, T. M. 1950. Beyond the gene—two years later. In *Science in progress*, ed. G. A. Baitsell, 167. 7th ser. New Haven, CT: Yale University Press.
46. Hertwig, R. 1914. Über parthenogenesis der Infusorien und die Depressionszustände der Protozoen. *Biol. Zentralbl.* 34:557.
47. Woodruff, L. L. and Erdmann, R. 1914. A normal periodic reorganization process without cell division in *Paramecium*. *J. Exp. Zool.* 17:425.
48. Sonneborn, T. M. 1939. Genetic evidence of autogamy in *Paramecium aurelia*. *Abstr. Anat. Rec.* 75(Suppl. 43):85.
49. Beale, G. 1952. Antigenic variation in *Paramecium aurelia*, variety 1. *Genetics* 37:67.
50. Sonneborn, T. M. 1937. The extent of the interendomictic interval in *P. aurelia*, and some factors determining its variability. *J. Exp. Zool.* 75:471.
51. Tsukii, Y. and Hiwatashi, K. 1979. Artificial induction of autogamy in *Paramecium caudatum*. *Genet. Res.* 34:163.
52. Mikami, K. and Hiwatashi, K. 1975. Macronuclear regeneration and cell division in *Paramecium caudatum*. *J. Protozool.* 22:537.
53. Sonneborn, T. M. and Schneller, M. V. 1979. A genetic system for alternative stable characteristics in genomically identical homozygous clones. *Dev. Genet.* 1:21.

54. Sonneborn, T. M. 1954. Is gene A active in the micronucleus of *Paramecium aurelia*? *Microb. Genet. Bull.* 11:25.
55. Wichterman, R. 1939. Cytogamy: A new sexual process in joined pairs of *Paramecium caudatum*. *Nature* 144:123.
56. Sonneborn, T. M. 1954. The relation of autogamy to senescence and rejuvenescence in *Paramecium aurelia*. *J. Protozool*. 7:38.
57. Sonneborn, T. M. 1975. *Paramecium aurelia*. In *Handbook of genetics*, ed. R. C. King, chap. 20. Vol. 2. New York: Plenum Press.
58. Beisson, J. and Capdeville, Y. 1966. Sur la nature possible des étapes de differenciation conduisant à l'autogamie chez *Paramecium aurelia*. *C. R. Acad. Sci*. 263:1258.
59. Aufderheide, K. J. 1987. Clonal aging in *Paramecium tetraurelia*. II. Evidence of functional changes in the macronucleus with age. *Mech. Aging Dev*. 37:265.
60. Gilley, D. and Blackburn, E. H. 1994. Lack of telomere shortening during senescence in *Paramecium*. *Proc. Natl. Acad. Sci. U.S.A*. 91:1955.
61. Takagi, Y. 1988. Aging. In *Paramecium*, ed. H.-D. Görtz, chap. 9. Berlin: Springer-Verlag.
62. Sonneborn, T. M. 1957. Breeding systems, reproductive methods, and species problems in protozoa. In *The species problem*, ed. E. Mayr, 155. Washington, D.C.: AAAS.
63. Dippell, R. V. 1955. A preliminary report on the chromosomal constitution of certain variety 4 races of *Paramecium aurelia*. *Caryologia Suppl*. 1109.
64. Sonneborn, T. M. and Dippell, R. V. 1960. Cellular changes with age in *Paramecium*. In *The biology of aging, American Institute of Biological Sciences, Symposium 6*, ed. B. L. Strehler, 285. Baltimore: Waverly Press.
65. Mitchison, N. A. 1955. Evidence against micronuclear mutations as the sole basis for death at fertilization in aged, and in the progeny of ultra-violet irradiated *Paramecium aurelia*. *Genetics* 40:61.
66. Sonneborn, T. M. 1954. Gene-controlled, aberrant nuclear behavior in *Paramecium aurelia*. *Microb. Genet. Bull*. 11:24.
67. Sonneborn, T. M. and Schneller, M. V. 1955. Genetic consequences of aging in variety 4 of *Paramecium aurelia*. *Rec. Genet. Soc. Am*. 24:596.
68. Siegel, R. W. 1963. New results on the genetics of mating types in *Paramecium bursaria*. *Genet. Res*. 4:132.
69. Preer Jr., J. R. and Preer, L. B. 1979. The size of macronuclear DNA, and its relationship to models for maintaining genic balance. *J. Protozool*. 26:1979.
70. Kimball, R. F. 1953. The structure of the macronucleus of *Paramecium aurelia*. *Proc. Natl. Acad. Sci. U.S.A*. 39:345.

4 Methods of Studying Genetic Processes in Organisms of the *Paramecium aurelia* Species Group

In contrast to other organisms that have been the basis of genetic studies in the past (*Drosophila*, peas, maize, *Neurospora*, *E. coli*, etc.), *Paramecium* is very unusual. It has both micronuclei and macronuclei, and a wealth of nuclear processes that are especially useful for genetic analysis, from conjugation to autogamy to macronuclear regeneration (see Chapter 3). As we will see, it also sometimes exchanges cytoplasm with its partner in conjugation, and it does so in amounts that can be measured. In spite of its peculiarities, it has many characteristics of other forms of life that can be used for studying genetics by standard methods. We will pay particular attention to some of these features that are especially characteristic of *Paramecium* and do not exist in higher organisms.

4.1 ISOLATION OF CELLS AND PREPARATION OF PURE CLONES

Establishment of pure clones of *Paramecium* is easily accomplished by isolating single cells from the wild from a mixed assortment of ciliates and other organisms in streams or ponds, by putting one of the cells provisionally identified as *Paramecium*, isolated with a micropipette, into a suitable artificial medium. A good culture medium consists of 1.5 g of dessicated cereal grass marketed under the name Cerophyl (Pines International Co., Lawrence, Kansas), 0.6 g of Na_2HPO_4 (anhydrous), and 1,000 µg of stigmasterol (0.4 ml of a stigmasterol solution made by dissolving 1 g of stigmasterol in 400 ml of acetone). With some batches of Cerophyl, growth rates and population densities may be improved by adding 0.1 g of yeast extract (Difco) to every liter of culture medium. Water is added to a total of 1 liter. The mixture is boiled for about 5 minutes, cooled, and filtered through absorbent cotton, autoclaved, and stored. The day before use it is inoculated with a suitable bacterial strain such as *Klebsiella aerogenes* or a strain that is found by trial and error to be capable of producing good growth of paramecia. The cells are washed through this medium by passing them through a series of samples of medium, and finally allowed to grow up into a pure culture free of other organisms, except for the strain of bacteria used for inoculation. Since each *Paramecium* is a single cell having a characteristic genotype, this process leads to the formation of pure clones that are

43

called stocks. Stocks are always derived from a single isolate from the wild and are usually homozygous because of the frequent occurrence of autogamy, which renders all genetic loci homozygous (see Chapter 3). Further operations can be carried out with one of the homozygous ex-autogamous clones thus prepared, after confirming that the originally isolated organism belongs to one of the species of the *P. aurelia* complex (see Chapter 2).

4.2 OBTAINING COMPLEMENTARY MATING TYPES

After the discovery of mating types in the *P. aurelia* complex by Sonneborn in 1937[1] (as described in Chapter 2), attempts were made to conduct genetic analysis by sexual recombination, as had been done previously with many other organisms in which the gametes and zygotes give rise directly to cells of the following generations, and zygotes are formed simply by fusion of gametes. Thus, the first step in the genetic analysis of a newly isolated sample of *Paramecium* is to obtain the two mating types of the strains being investigated, so that crosses involving sexual recombination can be carried out. In group A species, for example, *P. primaurelia* (see Chapter 5), cells of either mating type of a new isolate can easily be obtained and crossed with a standard strain, since both mating types are obtained in appreciable quantities after autogamy in these species.

In group B there is often little or no change of mating type following autogamy, and only one mating type of the new form may then be available for analysis. If it is desired to analyze the genetic basis of the new form, this could not therefore be done by crossing it with one of the mating types of a standard strain in group B, if only one mating type is available. Nevertheless, it is usually found that, in time (sometimes quite a long time), the two different mating types of any variant will eventually become available to an experimenter even in the group B species, because there are rare changes of mating type during conjugation or autogamy even in the B group (see Chapter 5). Both mating types can then be obtained, and the new isolate can be crossed with a standard stock. It is therefore theoretically possible to make an analysis of any variants of the *P. aurelia* complex by crosses.

4.3 MENDELIAN TRANSMISSION

The usual method of Mendelian analysis with *Paramecium* is to cross a newly isolated cell bearing a recognizable phenotype with a standard strain of the same species bearing an alternative phenotype, observe the nature of the F1s, and pass the latter through autogamy to obtain F2s and make a series of autogamous isolations. Autogamy can be obtained by growing the F1s for a sufficient number of fissions at the maximum rate after crossing, and then subjecting some of the F1 cells to a period of starvation, causing autogamy to take place. Autogamy can be shown to have occurred by staining a sample of cells with acetocarmine, and observing that macronuclear fragments and a newly forming anlage (macronucleus) are present in all stained cells.

Macronuclear regeneration in which macronuclear fragments regenerate into new macronuclei, instead of arising from fusion of a nucleus from each parent during conjugation (see Chapter 3), occurs with some frequency and produces a spurious F1 and F2 generation. Pairs arising from macronuclear regeneration can be detected by staining of a portion of the progeny in any line of descent at different numbers of fissions after autogamy and rejecting any lines that show that macronuclear fragments are regenerating rather than degrading. Alternatively, one may include a marker gene known to segregate properly in crosses. If the marker segregates properly in the F2 from a pair, macronuclear regeneration is eliminated. F2 cells are allowed to pass through at least five fissions and examined by testing them for the character being studied. Most genes require at least five fissions before they are fully expressed. In cases of straightforward Mendelian inheritance in strains with a single allelic difference, for example, such as temperature sensitivity (failure of paramecia to multiply at high temperature), the results are usually quite clear, the two homozygous "allelic" types appearing in approximately equal numbers in ex-autogamous F2s.

4.4 CARYONIDAL TRANSMISSION

To understand caryonidal inheritance we need first to define the caryonide.[1] Conjugation and autogamy have already been briefly described in Chapter 3, but will be reconsidered here in more detail. Such detail is necessary to understand caryonidal inheritance.

As already noted, conjugation of two animals produces four caryonides, while in autogamy, where the same process occurs in one animal, two caryonides are produced. Just before the first cell division late in the processes of autogamy and conjugation, each cell contains one fusion diploid micronucleus (see Figure 3.12, stage 7), a small open circle at the back of each cell. Next (stage 8), this fusion nucleus undergoes two mitotic divisions to produce in each cell four diploid nuclei, seen as two small open circles at the front and the back of each cell. Since these four nuclei were formed by mitosis from one nucleus, they should all four be identical. Now (stage 9) two of the four micronuclei in each cell enlarge and develop into macronuclear anlagen, while the other two remain micronuclei. In each dividing cell (stage 10) these newly forming macronuclei do not divide, but are segregated, one to each of the two new cells. The micronuclei divide mitotically and are distributed two to each new cell. At the end of this process (stage 11), after cell division, the macronuclear development is complete. Each one of these first four cells produced at conjugation, or two cells produced at autogamy, together with all of the future progeny of that cell, is called a caryonide. Note that autogamy produces two caryonides, just as does each of the two cells in conjugation. Since the process of macronuclear development is a complex one entailing many cuts and ligations in the micronuclear DNA, it might well be that events in the development of the new macronuclei are different. If so, then the differences in phenotypes produced by the two caryonides might be measurable and can be detected as caryonidal inheritance by the geneticist. Heritable differences between the two caryonides derived from a single cell are generally due to caryonidal inheritance, not cytoplasmic inheritance, for the two anlagen develop in the same cytoplasm.

Caryonidal inheritance was established early in the history of ciliate genetics, and although it was realized from the first that it indicated some sort of change in macronuclei when they were first formed, the fact that they are produced by rearrangements in the DNA was not known until much later, when changes could be followed at the molecular level. As we will see, they constitute examples of numerous changes that are attributable to modification in nuclei during their formation. Such changes have often been referred to as being epigenetic rather than Mendelian in nature. Although changes in DNA have not been demonstrated for all these examples, they have been in numerous cases. The first case of caryonidal inheritance was shown by the inheritance of mating type in the A group of species, where the mating types are distributed at random among the products of the first fission after autogamy or conjugation. This is, of course, the fission at which macronuclei derived from new anlagen are separated. At subsequent fissions they do not change. It appears that in this case the DNA in the anlagen becomes modified at the time of its formation to produce a new stable variant macronucleus. In the case of mating type, the epigenetic molecular mechanism has not been found.

In other cases, however, unlike the previous example, when the cytoplasm also plays a role in determining the character, the two caryonides are often alike, tending to hide the caryonidal inheritance. In this event, another technique is useful in distinguishing caryonidal inheritance from cytoplasmic inheritance, in other words, macronuclear inheritance from cytoplasmic inheritance—the heterocaryon test. See Figure 5.3. In the heterocaryon test cytoplasmic exchange and macronuclear regeneration are combined. Conjugants undergoing cytoplasmic exchange are picked by virtue of the fact that they show delayed separation of the conjugants, and macronuclear regeneration is induced by exposing the conjugants to high temperature at the time of anlagen production, which causes a delay in new macronuclear development. With suitably genetically marked strains it is easy to see that lines that develop from an old macronucleus always retain their previous mating type (and marker genes), and lines that develop from a new macronucleus often undergo change of their mating type (and marker genes). Nanney,[2] using data of Sonneborn,[3] points out that this test provides clear evidence for caryonidal inheritance of mating types in the case of *P. tetraurelia* and shows that the macronucleus, not the cytoplasm, ultimately controls the phenotype of mating types in *P. tetraurelia*.

When the lines tend to be unstable, it may also be difficult to recognize caryonidal inheritance. This effect is demonstrated by the behavioral mutant paranoiac that was analyzed by Rudman and Preer.[4] The key to recognizing it is to see that the character becomes more stable at the first fission after conjugation when the anlagen are segregated.

4.5 CYTOPLASMIC TRANSMISSION

In many cases traits fail to segregate in familiar Mendelian fashion, but instead follow the cytoplasm. In that case, we wish to identify the parental origin of the crossed cells by observing a suitable character at the "one fission" stage, after F1 exconjugants have divided once. One of the two cells from each exconjugant may be sacrificed to ascertain the phenotype. Since most (but not all) genes require four or five fissions to come to expression after they appear in a cell, the phenotype of the cell may be taken as that of the parent from which it was derived. The same result can also be obtained by checking the antigenic type of cells immediately after making a cross, because the characters at that time will be the same as the antigenic type of the parental cells and do not require an extra marker gene.

The characters that follow the cytoplasm are numerous and are based on a variety of mechanisms. They are classified as epigenetic mechanisms. In one group are the mitochondrial characters and symbionts like kappa that rely on their own DNA to be self-perpetuating. They are thought to be once free-living organisms that have been integrated into the protoplasm of the cell mitochondria a very long time ago, and kappa and its relatives much more recently.

Another series of examples are the cortical traits, in which newly forming structures are developed in strict relation to preexisting structures. Cortical traits also appear to follow the cytoplasm, but their determinants actually reside in the cortex of the cell (see Chapter 10), and thus are based on a very different mechanism from symbionts and mitochondria.

Serotype variants display still another basis for cytoplasmic heredity, relying on stable states of gene expression for their perpetuation. Stock 51, for example, has eleven or more genes each coding for a separate surface protein, but only one can be expressed at a time. These types tend to be relatively stable under specific environmental conditions. In fact, virtually all are stable when fed with enough food to multiply at one fission per day at 27°C. Serotypes are said to switch from one to the other when a locus turns off and another turns on (see Chapter 7). Crosses between the serotypes reveal that offspring are like their parents and thus show the cytoplasmic pattern of inheritance.

Another example of cytoplasmic inheritance with a still different mechanism is that of the d48 mutation found by Epstein and Forney[5] in *P. tetraurelia*. The d48 mutant cannot express the A serotype, but differs from most genic mutations that show normal Mendelian segregation, for it appeared to be inherited cytoplasmically. Epstein and Forney were able to show that the wild-type allele responsible for coding the A antigenic protein was present in the micronucleus, but missing in the macronucleus. The *A* gene was systematically deleted every time a new macronucleus was made at autogamy in *Paramecium*. Moreover, it was found that the wild type could be restored by injecting the *A* gene into the macronucleus of the mutant. Somehow the old macronucleus determines the character of the new macronucleus, and it is clear we are dealing with macronuclear determination instead of simple cytoplasmic inheritance. Injection of fluids and transplantation of nuclei, both micronuclei and macronuclei, have been important in working out the details of inheritance in d48. A macronuclear pattern of inheritance (from the old macronucleus to the new macronucleus) is produced, which is, of course, indistinguishable from the cytoplasmic pattern. However, d48 turns out to be an example of a much more general phenomenon, as will now be explained.

The story begins with that of gene silencing. Ruiz et al.[6] showed that the injection of virtually any gene from *P. tetraurelia* into wild-type organisms will silence the expression of that gene during subsequent vegetative fissions. The phenomenon is called posttranscriptional gene silencing (PTGS) and is highly specific for the gene injected. PTGS occurs only before autogamy, for after autogamy silencing disappears. Injections are made into wild-type vegetative cells in high copy number, and the effect is seen in those cells. It has also been found that if *E. coli*–bearing RNA copies of the gene are eaten by *Paramecium*, a very simple substitute for microinjection is provided; paramecia take up the gene copies and PTGS is produced. See also Chapter 12 for a detailed discussion of PTGS.

Garnier et al.[7] studied PTGS lines in subsequent generations after, rather than before, autogamy and found many stable mutant lines were induced, provided the genes in the recipient were not expressed and when the injection of genes was in sufficiently high copy number. The process was highly specific, for whatever DNA was introduced, deletions of those specific sequences were deleted from the new macronuclei. When crossed to wild type, these lines were found to show the cytoplasmic pattern of inheritance. The process was called homology-dependent inheritance. Moreover, introduction of the *A* gene could, under the proper circumstances, induce d48-like mutants that are indistinguishable from the original d48. So d48 is evidently a special case of this phenomenon. The fact that such inherited changes can be found in virtually any part of the genome is, of course, highly significant. That such events can be induced suggests that they might occur naturally. If so, they could account for many of the cases of non-Mendelian genetics in *Paramecium* genetics. Final elucidation of the molecular mechanisms producing these phenomena will be of the greatest importance in understanding ciliate genetics. Sonneborn would have liked these results, for he always said that the secret of *Paramecium* genetics lies in the macronucleus. See Chapter 12 for further discussion of homology-dependent inheritance.

4.6 METHODS OF IDENTIFYING GENETICALLY DEFINED GENES WITH THEIR DNA

The genomes of numerous organisms have now been completely sequenced, including that of *P. tetraurelia* strain d4-2. A complete library of the macronuclear genome now exists. See Chapter 8. The library can be searched with DNA sequences (http://paramecium.cgm.cnrs-gif.fr/blast). Many techniques have been devised to identify genetically defined genes with the DNA in a library that constitutes those genes. We will now try to elucidate several examples.

4.6.1 IDENTIFICATION OF THE RNAs PRODUCED BY GENES

If the messenger RNA of a gene can be obtained, the mRNA can be made radioactive and the library probed with the radioactive label to search for homologous sequences. This is possible for the very large serotype protein molecules whose mRNA can be easily identified on gels because of their size and abundance, radioactively labeled, and used as probes of a library. An example is the group of surface antigens whose mRNAs were discovered[8] and isolated. The genes were then easily found by probing a macronuclear library.[9] A similar process has been used to discover rDNA genes, distinguished by the fact that their RNA transcripts are present in the genome in very high numbers.

4.6.2 IDENTIFICATION BASED ON PROTEIN SEQUENCE

If the amino acid sequences in a protein can be discovered, then it is sometimes possible to synthesize a fragment of RNA or DNA that will hybridize with the gene. The procedure is complicated by the fact that several different codons may be used by an organism for one particular amino acid. However, by reference to the coding preferences of a given organism and by creating ambiguities in the code, such searches are usually successful. An example is the CAM gene coding for calmodulin in *P. tetraurelia*.[10]

4.6.3 IDENTIFICATION BASED ON SIMILARITY OF PROTEINS

Proteins in related species are often sufficiently like each other so that a probe made to one whose sequence is known will be sufficiently like that in the other species to hybridize and produce a signal when a genomic library is searched. Alternatively, if the complete genome has been sequenced, then DNAs that are similar will be returned by a computer search of the genome using the sequence of the related protein.

4.6.4 FUNCTIONAL COMPLEMENTATION

When any DNA is injected into the macronucleus of *Paramecium*, it becomes linearized, telomeres are added to the ends, and it will multiply and genes of the injected DNA are expressed.[11,12] If a wild-type gene has been cloned and is injected into a mutant that is defective in that gene, the mutant will often be "cured" and will become wild type until the effect is lost at the next autogamy when a new macronucleus is formed. When injections are made, they can be quite large, even amounting to the quantity of DNA in a whole macronucleus. Haynes et al.[13] injected a replicate of a whole library into a mutant and found that the gene in the library was present in enough copies to bring about the cure of the specific behavioral mutant. Using this technique he was then able to try to select a single clone from the library that would bring about complementation when injected into the mutant. He first plated out the library onto several plates with five thousand colonies per plate and prepared replicas of each plate. Injection of the DNA from each of these plates revealed one that had the gene and was able to complement the mutant. A replica of this plate was further divided into subsections and the subsections were tested for activity. Continuation of this process eventually led to the single colony on the original plate that was complementing. Keller and Cohen[14] made the process much simpler by transferring 60,000 clones of a library to 60,000 numbered wells in 384-well microtiter plates (160 plates). Once this initial work was done, selection for the individual clone became much easier and more reliable. The method depends only on the ability of an injected wild-type library to cure a mutant defective in its wild-type activity. The method does not require that the gene product be known.

4.6.5 MALDI-TOF ANALYSIS OF *PARAMECIUM* PROTEINS

MALDI-TOF is an acronym for matrix-assisted laser desorption/ionization time-of-flight analysis.[15] The method utilizes a mass spectrometer. One starts with a small sample of test protein of interest that has been isolated from a gel (absolute purity is not required). The test protein is then digested with trypsin to cut it into peptides at arginine and lysine residues in the protein sequence. The peptides are now analyzed by MALDI-TOF to produce a spectrum of peptides of very precise molecular weights. The success of the method is dependent on being able to predict these same molecular weights from a DNA sequence. The library is searched for DNA sequences that can produce mRNAs that when translated into protein and cut at arginine and lysine, produce molecular weights like those produced by the protein in question and determined by MALDI-TOF analysis. The method, of course, is dependent on having the whole genome of *Paramecium* available, and the use of a specialized computer program. The complete genomic sequence is now available for *P. tetraurelia* on the Web. It is assumed that the programs for doing the search will be available soon.

REFERENCES

1. Sonneborn, T. M. 1937. Sex, sex inheritance and sex determination in *Paramecium aurelia*. *Proc. Natl. Acad. Sci. U.S.A.* 23:378.
2. Nanney, D. L. 1957. Mating type inheritance at conjugation in variety 4 of *Paramecium aurelia*. *J. Protozool.* 4:89.
3. Sonneborn, T. M. 1954. Patterns of nucleo-cytoplasmic integration in *Paramecium*. *Caryologia Suppl.* 6:307.
4. Rudman, B. M. and Preer Jr., J. R. 1996. Non-Mendelian inheritance of revertants of *paranoiac* in macronuclear development in *Paramecium*. *Genetics* 129:47.
5. Epstein, L. M. and Forney, J. D. 1984. Mendelian and non-Mendelian mutations affecting surface antigen expression in *Paramecium tetraurelia*. *Mol. Cell. Biol.* 4:1583.
6. Ruiz, F., et al. 1998. Homology-dependent gene silencing in *Paramecium*. *Mol. Biol. Cell* 9:931.
7. Garnier, O., et al. 2004. RNA-mediated programming of developmental genome rearrangements in *Paramecium tetraurelia*. *Mol. Cell. Biol.* 24:7370.
8. Preer Jr., J. R., Preer, L. B., and Rudman, B. M. 1981. mRNAs for the immobilization antigens of *Paramecium*. *Proc. Natl. Acad. Sci. U.S.A.* 78:6776.
9. Forney, J. D., et al. 1983. Structure and expression of genes for surface proteins in *Paramecium*. *Mol. Cell. Biol.* 3:466.
10. Kink, J. A., et al. 1990. Mutations in *Paramecium* calmodulin indicate functional differences between the C-terminal and N-terminal lobes *in vivo*. *Cell* 62:165.
11. Godiska, R., et al. 1987. Transformation of *Paramecium* by microinjection of a cloned serotype gene. *Proc. Natl. Acad. Sci. U.S.A.* 84:7590.
12. Gilley, D., et al. 1988. Autonomous replication and addition of telomere-like sequences to DNA injected into *Paramecium* macronuclei. *Mol. Cell. Biol.* 8:4765.
13. Haynes, W. J., et al. 1998. The cloning by complementation of the *pawn-A* gene in *Paramecium*. *Genetics* 149:947.
14. Keller, A.-M. and Cohen, J. 2000. An indexed genomic library for *Paramecium* complementation cloning. *J. Eukaryotic Microbiol.* 47:1.
15. Henkin, J. A., et al. 2004. Mass processing—An improved technique for protein identification with mass spectrometry data. *J. Biomol. Technol.* 15:230.

5 The Determination of Mating Types in *Paramecium*

5.1 INTRODUCTION

The mating types of the *P. aurelia* complex of species[1] originally were designated in *P. primaurelia* mating types I and II, *P. biaurelia* mating types III and IV, *P. triaurelia* mating types V and VI, and so on. Each of the fourteen species of the *P. aurelia* complex of species has two mating types, one called O (odd) and the other E (even). Moreover, there is no exception to the rule that the O mating types are all homologous to each other, as are the E mating types. The homologies depend on several factors: the time it takes for a new genotype to be expressed (phenomic lag), response to temperature, and cross-reactions. Cross-reactions occur only between the O of some species and the E of another (see Table 2.2).

The species can be grouped by considering their mode of mating type inheritance. The species in group A (species 1, 3, 5, 9, 11, and 14) show caryonidal inheritance and no cytoplasmic effect. The species of group B (species 2, 4, 6, 7, 8, 10, and 12) show cytoplasmic inheritance. In group C inheritance is Mendelian. As we will see, group B species also show a weak caryonidal effect. The inheritance of mating types will now be considered in more detail.

5.2 CARYONIDAL INHERITANCE

Most of the stocks in the group A species are able to produce both complementary mating types, that is, are two-type stocks, with mating types O and E. These are contrasted with the one-type stocks, producing only type O. One-type stocks, of course, cannot mate within the stock. Analysis of the progeny produced by pairings of the opposite mating types within a two-type stock in the A group of species always reveals caryonidal inheritance, with no effect of the parental cytoplasm on the mating type of the next generation. Thus, each exconjugant and each autogamous cell has two new macronuclei formed within it, producing two caryonides when the new macronuclei are segregated to individual cells at the first cell division. The new mating type is set by the newly formed macronucleus in each of the two cells, independently of each other and independently of the parental cytoplasm, and this mating type remains constant throughout subsequent generations of asexual reproduction. However, at the next sexual reproduction, at autogamy or conjugation, the macronucleus breaks up and is lost. For example, in *P. primaurelia* the two individual caryonides coming from each cell may both be type O, both be type E, or one

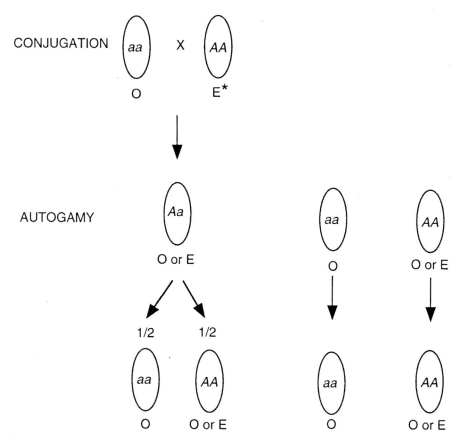

FIGURE 5.1 A one-type stock of genotype *aa* and phenotype mating type O is mated to a two-type stock whose genotype is *AA* and phenotype E.* After conjugation, because *A* is dominant to *a*, the phenotype of the progeny of the cross is genotype *Aa* and mating type O or E. After autogamy half the clones become *aa* (type O) and the other half *AA* (type O or E). Autogamy in the one-type stock produces all clones *aa* (type O), and autogamy in the two-type stock produces all clones *AA* (type O or E).
*Although *AA* can produce either mating type O or E, here it must be E in order for it to conjugate with O.

may be type O and the other type E. There is no correlation between the new mating type and the old mating type of the cell in which the two anlagen are formed in *P. primaurelia*, nor in any of the other stocks found in the group A species. Sonneborn[2] showed that this determination is dependent at least in part on external factors. Thus, he demonstrated that in *P. primaurelia* the frequency of mating type E, measured as a percent, is linearly related to the temperature at the time of formation of the caryonide, varying from 38% at 10°C to 88% at 35°C in one stock.

A cross of a one-type stock with a two-type stock produces in the F1 after autogamy a 1:1 segregation of one-type versus two-type (see Figure 5.1; Sonneborn[3]; Wichterman[4]). In other words, segregation is strictly Mendelian. This was the first gene found in protozoa.[3] Butzel[5] derived a series of mutants at one locus from a

The Determination of Mating Types in *Paramecium*

two-type stock in *P. primaurelia* that acted exactly like one-type stocks in crosses. Since he could find mutants that became pure for mating type O, but no mutants that were pure for mating type E, he conjectured that the mating type O substance was produced by the cell first and was subsequently converted to the mating type E substance. Although this still remains a viable hypothesis, no definitive evidence has ever been provided.

Is it possible that structures other than macronuclei also segregate at the first fission after conjugation, and determine mating type instead of the new macronuclei? No, further evidence clearly rules out this possibility. Sonneborn[6] showed that when multiple anlagen are sometimes produced and segregate at later fissions, both caryonides and mating types segregate together. Moreover, at macronuclear regeneration there is no change in mating type. The fragments of the old macronucleus are, of course, already determined for mating type and remain so. *P. multimicronucleatum* also shows caryonidal inheritance of mating types after conjugation, but in this case four rather than two anlagen are produced in each exconjugant. Segregation of mating types and caryonides in this species does not occur until the second fission after conjugation, rather than the first fission as in the *P. aurelia* group.

5.3 THE CYTOPLASMIC EFFECT

Inheritance in the group B species (species 2, 4, 6, 7, 8, 10, and 12) is determined primarily by the cytoplasm, exconjugants producing stable lines of the same mating type as their cytoplasmic parent. Once set, the mating type never changes during subsequent fissions. Cytoplasmic exchange, however, does result in mating type change. See Figure 5.2.

Nevertheless, it is clear that simple cytoplasmic inheritance is not an adequate explanation for inheritance in the group B species, for experiments show that mating type is determined by the macronucleus in the B group (as well as in group A). The cytoplasm of the O mating type determines the newly forming macronuclear anlagen to be O determining, and the cytoplasm of mating type E cells determines the anlagen of mating type E to be E determining (see Beale[7]). Once set, the macronucleus maintains its mating type for the life of the macronucleus until the next autogamy or conjugation occurs.

A hint of this nuclear determination was provided by Nanney.[8] In a study involving selfers (unstable mating type lines—see section below) he reported caryonidal inheritance in stock 51, *P. tetraurelia*, of group B. Caryonidal inheritance, of course, implies nuclear determination. He found that occasionally the two caryonides from an exconjugant produced sister caryonides, one of which was O and the other E, showing caryonidal inheritance. These types were very stable after they were formed, with no change occurring in later vegetative fissions.

Proof in the group B species that mating type is actually determined by the anlagen and the macronuclei that they produce was provided by Sonneborn[9] (see Figure 5.3). First, he noted that when cytoplasmic exchange occurred in stock 51 of species *P. tetraurelia*, the mating type O partner usually became E. In a cross of mating type O with mating type E he induced both cytoplasmic exchange and macronuclear regeneration in the presence of marker genes. Both anlagen and old macronuclear fragments survived, and the only cells that showed change to a new

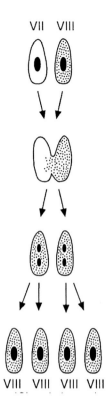

FIGURE 5.2 Apparent cytoplasmic inheritance of mating type in the group B species of the *P. aurelia* complex. Reprinted with permission of Cambridge University Press; modified from Beale, G. H., *The Genetics of* Paramecium aurelia (London: Cambridge University Press, 1954), 133, Figure 10.

mating type came from anlagen, while all cells derived from old macronuclear fragments retained their original mating type.

Byrne[10] treated stock 51 mating type E with mutagens and produced several mutants that were stable O. When these mutants were crossed back to wild type, the progeny showed typical Mendelian inheritance. Three unlinked loci, able to inhibit the expression of mating type E and produce stable mating type O, were discovered, and others likely exist. These were called mt^A, mt^B, and mt^C. In more recent nomenclature these might be designated mtA^O, mtB^O, and mtC^O. Wild type that can express either O or E would be designated either $mtD^{O,E}$ or mtD^+. It is interesting to note that several separate loci can affect mating type determination.

5.4 INDEPENDENCE OF THE CYTOPLASMIC EFFECT AND THE PHENOTYPE

That a cell might be of one mating type and yet have a hidden cytoplasmic state that might be another was first discovered by Nanney[8,11] in a study of selfing (see Section 5.8) in *P. tetraurelia*. He often found mating type E selfers of stock 51 (*P. tetraurelia*,

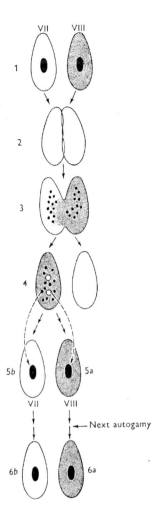

FIGURE 5.3 Old macronuclear fragments never change hereditary determinants. Induction of macronuclear regeneration and cytoplasmic exchange never produce a change in nuclei derived from old macronuclei, but nuclei derived from anlagen often change. Reprinted with permission of Cambridge University Press; modified from Beale, G. H., *The Genetics of Paramecium aurelia* (London: Cambridge University Press, 1954), 135, Figure 11.

group B) that produced the complementary type at such a low rate they were difficult to identify as selfers. However, when these E selfers underwent autogamy, they proved to have the cytoplasmic state of O, for they produced almost invariably 100% O. These results show that the cytoplasmic determinant of mating type can be independent of the phenotype. The mating type E individuals that he started with had a hidden cytoplasmic state of mating type O, and it was expressed after autogamy.

A similar effect is seen in the experiments of Byrne[10] on stock 51 of *P. tetraurelia*. Byrne attempted to get his mutants by mutagenizing mating type E cells and looking for the presence of mating type O cells. His mutants, although phenotypically

O, had a hidden cytoplasmic state that remained E. When a wild-type gene was introduced back into the O mutant by means of a cross, the mating type of the cell always changed back to the type specified by the hidden cytoplasmic state, which had remained constant, even though the mutation had forced the mating type to be different. Although these mutants affect the phenotype, they do not induce changes in the hidden cytoplasmic state.

Taub[12–14] also showed separation of mating type from the cytoplasmic state of mating type. Taub showed that stocks 227 and 253 (*P. septaurelia*, group B) demonstrate the typical cytoplasmic inheritance expected of group B two-type stocks. When mating type O was crossed with mating type E of these two stocks there was no change in mating type in the F1, nor in subsequent generations, with the O remaining O and the E remaining E. However, Taub did show that stock 38 (*P. septaurelia*) differed from these two stocks in that it had a naturally occurring allele that caused its hidden cytoplasmic state to be E determining, even though its phenotype was mating type O. Thus, a cross of stock 38 mating type O (with a hidden cytoplasmic state of E) with stock 227 E or stock 253 E produced in the F1 a heterozygous genotype. This genotype allowed the hidden mating type of E to appear, both F1 exconjugants producing all mating type E cells. Other crosses show that stock 38 differs from Byrne's case (in group B, stock 51) in that it induces the hidden cytoplasmic state itself to "mutate" and to become E determining. Another gene at a second locus was produced by a mutation induced by ultraviolet light in a line containing the mating type gene found in stocks 227 and 253. The mutant is also mating type O with a hidden cytoplasmic state of mating type E, but Taub showed that, like Byrne's mutants, it had no mutagenic effect on the cytoplasmic state.

5.5 THE O* PHENOTYPE: A NEW CYTOPLASMIC STATE

The O* phenotype was discovered by Brygoo[15] and Brygoo and Keller.[16,17] In crosses of O by E in stock 51 and O by E in stock 32, normal cytoplasmic inheritance is obtained. However, in the cross of 32 E by 51 O, exceptional results are seen because an appreciable proportion of the heterozygous F1 lines derived from the E cytoplasmic member of the cross expresses a new cytoplasmic state, which he called O*. When these F1 heterozygotes are carried through autogamy to produce an F2, a 1:1 ratio of O:E is found. This result indicates segregation at a single locus that he calls *mtD*. The two alleles at the *mtD* locus, he states, are segregated into the genotypes mtD^{51}/mtD^{51} (type O*) and mtD^{32}/mtD^{32} (type E). He maintains that the cytoplasmic factor O* is different from the cytoplasmic factor O found in stock 51, for the weaker O* cytoplasmic state allows one to see the segregation of the mating type genes in the two stocks. The reader is referred to papers by Brygoo for further information about the O* lines.

5.6 THE PURE E MUTANT

As already pointed out, there are several mutants in *P. tetraurelia* that are always pure for mating type O. In addition, Brygoo and Keller[16,17] discovered a mutant in *P. tetraurelia* that was restricted to mating type E rather than O. It is called mtF^E.

The Determination of Mating Types in *Paramecium*

This mutant was unlike all others discovered in *P. tetraurelia* (and in *P. septaurelia*) in that it produced not O but E cells. mtF^E is not allelic with mt^A, mt^B, mt^C, or mt^D. Homozygotes for the gene have a cytoplasm that is always E determining, irrespective of whether it comes from an O or an E cell. Surprisingly, however, Brygoo and Keller found that mtF^E/mtF^E in the O* cytoplasm is always pure mating type O. The mtF^E homozygotes proved to be pleiotropic and very weak and difficult to culture.

Meyer and Keller[18] showed that in the mtF^E mutant there was a failure to excise an IES found in the serotype G gene of stock 51. This finding of a relationship between mating type and serotypes was quite surprising and will be discussed at the end of this chapter.

5.7 MICRONUCLEAR DETERMINATION

Working with stock 51 of *P. tetraurelia*, Sonneborn[8] produced evidence that the micronuclei, as well as the macronuclei and the cytoplasm, play a role in determining the mating type of the new macronuclei formed at conjugation and autogamy. These results were obtained by crossing normal lines to amicronucleate lines. Later Brygoo and Keller made additional crosses. Both the old and new data agreed remarkably well and were presented in a joint paper.[19] First they found in control crosses consisting of normal nucleated mating type E and normal mating type O that 94 to 99% of the exconjugants produced F1 exconjugants that showed no change in type. When the cross was made between a cell that lacked micronuclei and a cell that had normal micronuclei, the cell that lacked the micronuclei was destabilized. If the cell lacking micronuclei was mating type O, the cell produced only about 12 to 25% mating type O. If the cell lacking micronuclei was mating type E, then the cell produced about 77% mating type E. It is clear from these results that the micronuclei contribute to the determination of the mating phenotype, and indeed are partially predetermined. One nice cross that was made was a cross of mating type O* with mating type O. The cross was induced by adding reactive, isolated cilia, which are known to induce mating in cells of the same mating type.[20] Using this technique, Metz found that the two F1 exconjugants, which had the same genes, were nevertheless different, confirming that the cytoplasms were different.

5.8 SELFERS

Kimball[21,22] studied the small percentage of cells produced in all two-type stocks of *P. primaurelia* that are unstable and are classified as selfers. Kimball was able to demonstrate that in such selfing lines, the cells forming pairs do have mating types. He did this by splitting pairs with a micropipette and then testing them with mating type tester individuals. Both members of split pairs react with only one of the two testers, and if one member reacts with one tester, the other reacts with the other.

One kind of selfing is understood, that which is due to the variable lag in the expression of genes that have changed their determination and show phenotypic lag at conjugation and autogamy. When mating type E changes to mating type O (but not when O changes to E), phenotypic lag lasts for a few fissions. Due to asynchrony in the process, for a brief period both mating types are present and selfing occurs.[21,22] This same process was shown in *P. triaurelia* by Sonneborn.[23]

The major mechanism for selfing is still unknown. One possibility is that there are macronuclear units, some determined for one mating type, and others determined for the complementary mating type. During amitosis in the macronuclei, unequal divisions occur, and cells are produced on either side of the threshold and are then able to mate with each other. However, tests and data to support this hypothesis have not been forthcoming in any decisive fashion. For more detail on these tests, see Nanney,[24] Nanney and Caughy,[25] and Preer.[26]

5.9 MATING TYPES IN SPECIES OTHER THAN THOSE IN THE *P. AURELIA* COMPLEX

In addition to the *P. aurelia* group of species, mating types have been studied in other species of *Paramecium*, namely, *P. bursaria*, *P. caudatum*, and *P. multimicronucleatum*. Although much less is known about the genetic basis of the mating type systems of these species than of those in the *P. aurelia* complex, it is of interest to consider some of them. A few facts about mating types in these three species are therefore given in the following sections.

5.9.1 *P. BURSARIA*

Mating types in *P. bursaria* were first investigated by Jennings,[27,28] later by Siegel and Larison[29] and Siegel,[30] and finally by Bomford.[31,32] Six syngens (the term *syngen* has remained in use for *P. bursaria*) have been found, and each contains four or eight mating types, unlike the *P. aurelia* group of species, where there are only two mating types per species. In different syngens of *P. bursaria*, conjugation can occur between any two cells belonging to a different mating type in the same syngen. In syngen 1, which contains four mating types (denoted I, II, III, and IV) there are genes at two loci, *A* and *B*, with two alleles at each; that is, mating type I is determined by the dominant alleles *A* and *B*, type III by the double recessive *aa/bb*, and types II and IV by *aa/B–* and *A–/bb*, respectively. The notation *A–* or *B–* indicates presence of either one or two doses of a dominant allele.[29,30]

Bomford[31] made a provisional analysis of the genetics of syngen 3 of *P. bursaria* containing eight mating types, and found evidence for three dominant alleles at separate loci. Therefore, in this syngen there is direct genic determination of mating type, and no evidence of caryonidal or cytoplasmic determination like that found in the *P. aurelia* complex.

5.9.2 *P. CAUDATUM*

In general, the genetics of mating types in *P. caudatum* (in which use of the term *syngen* has also been retained) has been investigated by Hiwatashi in four of the known syngens, 1, 3, 12, and 13.[33–35] Analysis is more difficult than in the *P. aurelia* complex of species. First of all, there is no autogamy in *P. caudatum*, so that one cannot assume homozygosity at all loci, unless special steps are taken. Also, at conjugation nuclei are often not exchanged (cytogamy takes place) or, very frequently, macronuclear regeneration occurs. Moreover, at conjugation there are four, not two, new macronuclei, and hence an extra division is required before new caryonides can be segregated. Often when cells are mixed, pairing occurs between cells of the same mating type in the mixture. Marker genes are much fewer and often not employed. It appears, however, that simple genes appear to control the mating types, much like inheritance in species 13 of the *P. aurelia* group. For example, in syngens 3 and 12 of *P. caudatum* there are two mating types, denoted types O and E, with a dominant allele, mt^+, determining mating type E and a recessive allele, mt^-, determining mating type O.

Still, it is certain that epigenetic phenomena also occur, for selfing and change of mating type from one type to the alternative type that remain stable for many fissions can be demonstrated in some lines. Often selfing can be induced by a change in the growth conditions. In *P. caudatum*, the behavior of mating types following intersyngen crosses is remarkably different from that of intersyngen crosses in the *P. aurelia* group of species, which are generally infertile or lead to death. In *P. caudatum*, crosses between syngens can be produced by a special technique as described by Tsukii and Hiwatashi.[35] In one syngen of this species, when cells belonging to opposite mating types are mixed, conjugation occurs not only between the cells of different mating type (as in the *P. aurelia* complex), but also between cells of a single type, which are induced to undergo selfing, due apparently to the close proximity of regular pairing between different mating types. Moreover, when mixtures are made involving the two mating types of one syngen and those of another syngen, intersyngen hybrids are frequently produced and yield highly fertile progeny, a phenomenon that rarely, if ever, occurs in *P. aurelia*. After crosses involving syngens 3, 12, and 13 of *P. caudatum* were made, genetic analysis of the progeny of the intersyngen hybrids led to the conclusion that the mating type differences in *P. caudatum* were controlled by three genetic loci: one locus, *mt*, controlling the mating type differences within syngens, and two other loci concerned with intersyngen differences. Mating types in *P. caudatum* are thought generally to be under direct genic control, like that of species 13 of the *P. aurelia* complex, with the caveat that selfers of unknown significance exist.

5.9.3 *P. MULTIMICRONUCLEATUM*

There are two mating types per syngen, like the two mating types per species in the *P. aurelia* complex, but it is remarkable that certain stocks show a circadian rhythm involving mating type switches. In certain stocks of *P. multimicronucleatum*, all the cells exhibit one mating type during part of the day or night, and the other mating type during another part of the day or night. Hence, there are two switches during a 24-hour period, one from O to E and the other from E to O.[36] Selfing may occur at the time when the cells are switching over from one mating type to the other. The detailed times of switching vary according to the stock and the hours of light and darkness.[37] The periodic oscillation continues for a while during continuous light or darkness. However, the phenomenon seems to be initiated by the regular cycles of light and darkness and was found to occur only in stocks bearing a certain dominant gene, *C*. In the presence of the recessive allele, *c*, the cultures remain stable for one or the other of the mating types, and in that condition, inheritance of mating type is by the caryonidal system like that in group A species of *P. aurelia*.

There are numerous details about the switching phenomenon that are not reported here. The reader is referred to Barnett[37] and Karakashian[38] for more details about the circadian clocks in *Paramecium*.

5.10 A HYPOTHESIS TO EXPLAIN MATING TYPE INHERITANCE

It may be that mating type inheritance can be understood in terms of another totally different phenomenon, homology-dependent inheritance (discussed briefly in Chapter 4 and in greater detail in Chapter 12). In homology-dependent inheritance in wild type, a substance produced by the old macronucleus determines the newly forming macronucleus, to produce a normal undeleted nucleus that is like the old nucleus. If in a mutant the old macronucleus fails to produce this substance, then a deletion occurs and the mutant phenotype is produced.

As already pointed out, Meyer and Keller[18] (Section 5.6 above) showed that the *mtF^E* gene failed to excise a gene for serotype G. (They also produced a hypothesis much like the one presented here.) Moreover, they showed that this deletion was inherited exactly like a homology-dependent deletion. Although the relation to mating type is not explained in this case, the following considerations may help. Since it is estimated that there are many thousands of IESs in *Paramecium*, it seems unlikely that all of them are unique. Thus, it may be that many of the IESs that are removed from the genome are present in more than one gene. For instance, it could be that just by chance the IES that is cut out of the serotype gene is also cut out of the mating type gene, explaining the relationship between IES removal and mating type. Since the mating type gene has never been isolated biochemically, we do not know for sure whether sections of DNA are cut out. If whether a cell produces mating type O or E is dependent on whether an IES is removed, then mating type determination becomes a special case of homology-dependent mutation. The substance that passes from the old macronucleus to the new macronucleus in the case of homology-dependent inheritance has been shown to be nucleic acid (see Chapter 12). Therefore, one might assume that the same is true for mating type in the group B species.

On the other hand, if the deleted portion had no effect on development of the new macronucleus, and the deletions occurred spontaneously, then the new macronucleus would be independently determined and mating type inheritance would be like that in the group A species, that is, show caryonidal inheritance.

REFERENCES

1. Sonneborn, T. M. 1975. *Paramecium aurelia*. In *Handbook of genetics*, ed. R. C. King, chap. 20. Vol. 2. New York: Plenum Press.
2. Sonneborn, T. M. 1947. Recent advances in the genetics of *Paramecium* and *Euplotes*. *Adv. Genet.* 1:263.
3. Sonneborn, T. M. 1937. Sex, sex inheritance and sex determination in *P. aurelia*. *Proc. Natl. Acad. Sci. U.S.A.* 23:378.
4. Wichterman, R. 1985. *The biology of* Paramecium. 2nd ed. New York: Plenum Press.

5. Butzel, H. M. 1955. Mating type mutations in variety 1 of *Paramecium aurelia* and their bearing upon the problem of mating type determination. *Genetics* 40:321.
6. Sonneborn, T. M. 1938. Sex behavior, sex determination and the inheritance of sex in fission and conjugation in *Paramecium aurelia*. *Genetics* 23:169.
7. Beale, G. H. 1954. *The genetics of* Paramecium aurelia. London: Cambridge University Press.
8. Nanney, D. L. 1954. Mating type determination in *Paramecium aurelia*, a study in cellular heredity. In *Sex in microorganisms*, 266. Washington, D.C.: AAAS.
9. Sonneborn, T. M. 1954. Patterns of nucleocytoplasmic integration in *Paramecium*. *Caryologia* 6(Suppl.):307.
10. Byrne, B. 1973. Mutational analysis of mating type inheritance in syngen 4 of *Paramecium aurelia*. *Genetics* 74:63.
11. Nanney, D. L. 1957. Mating type inheritance at conjugation in variety 4 of *Paramecium aurelia*. *J. Protozool.* 4:89.
12. Taub, S. R. 1963. The genetic control of mating type differentiation in *Paramecium*. *Genetics* 48:815.
13. Taub, S. R. 1966. Regular changes in mating type composition in selfing cultures and in mating type potentiality in selfing caryonides of *Paramecium aurelia*. *Genetics* 54:173.
14. Taub, S. R. 1966. Unidirectional mating type changes in individual cells from selfing cultures of *Paramecium aurelia*. *J. Exp. Zool.* 163:141.
15. Brygoo, Y. 1977. Genetic analysis of mating-type differentiation in *Paramecium tetraurelia*. *Genetics* 87:633.
16. Brygoo, Y. and Keller, A.-M. 1981. Genetic determination of mating type differentiation in *Paramecium tetraurelia*. III. A mutation restricted to mating type E and affecting the determination of mating type. *Dev. Genet.* 2:13.
17. Brygoo, Y. and Keller, A.-M. 1981. A mutation with pleiotropic effects on macronuclearly differentiated functions in *Paramecium tetraurelia*. *Dev. Genet.* 2:23.
18. Meyer, E. and Keller, A.-M. 1996. A Mendelian mutation affecting mating-type determination also affects developmental genomic rearrangements in *Paramecium tetraurelia*. *Genetics* 143:191.
19. Brygoo, Y., et al. 1980. Genetic analysis of mating type differentiation in *Paramecium tetraurelia*. II. Role of the micronuclei in mating type determination. *Genetics* 94:951.
20. Metz, C. B. 1955. Mating substances and the physiology of fertilization in ciliates. In *Sex in microorganisms* [Symposium], ed. D. H. Wenrich, 284. Washington, D.C.: AAAS.
21. Kimball, R. F. 1939. A delayed change of phenotype following a change of genotype in *Paramecium aurelia*. *Genetics* 234:49.
22. Kimball, R. F. 1943. Mating types in the ciliate *Protozoa*. *Q. Rev. Biol.* 18:30.
23. Sonneborn, T. M. 1939. *P. aurelia*: Mating types and groups; lethal interactions; determination and inheritance. *Am. Nat.* 73:390.
24. Nanney, D. L. 1953. Mating type determination in *Paramecium aurelia*, a model of nucleo-cytoplasmic interaction. *Proc. Natl. Acad. Sci. U.S.A.* 39:113.
25. Nanney, D. L. and Caughey, P. A. 1955. An unstable nuclear condition in *Tetrahymena pyriformis*. *Genetics* 40, 388.
26. Preer Jr., J. R. 1968. Genetics of the protozoa. In *Research in protozoology*, ed. T. T. Chen, 129. Vol. 3. New York: Pergamon Press.
27. Jennings, H. S. 1941. Genetics of *P. bursaria*. II. Self-differentiation and self-fertilization of clones. *Proc. Am. Phil. Soc.* 85:25.
28. Jennings, H. S. 1942. Genetics of *Paramecium bursaria*. III. Inheritance of mating type, in cross and in clonal self-fertilization. *Genetics* 27:193.
29. Siegel, R. W. and Larison, L. L. 1960. The genetic control of mating types in *Paramecium bursaria*. *Proc. Natl. Acad. Sci. U.S.A.* 46:344.

30. Siegel, R. W. 1963. New results on the genetics of mating types in *Paramecium bursaria*. *Genet. Res.* 4:132.
31. Bomford, R. 1965. Changes in mating type in *Paramecium*. In *Proceedings of the 2nd International Congress on Protozool*, 251. Vol. 91. London: Excerpta Medica.
32. Bomford, R. 1966. The syngens of *Paramecium bursaria*: New mating types and intersyngenic mating reactions. *J. Protozool.* 13:497.
33. Hiwatashi, K. 1960. Analysis of the changes of mating type during vegetative reproduction in *Paramecium caudatum*. *Jpn. J. Genet.* 35:213.
34. Hiwatashi, K. 1968. Determination and inheritance of mating type in *Paramecium caudatum*. *Genetics* 58:373.
35. Tsukii, Y. and Hiwatashi, K. 1983. Genes controlling mating type specificity in *Paramecium caudatum*: Three loci revealed by intersyngenic crosses. *Genetics* 104:41.
36. Sonneborn, T. M. 1957. Diurnal change of mating type in *Paramecium*. *Anat. Rec.* 128:626.
37. Barnett, A. 1966. A circadian rhythm of mating type reversals in *Paramecium multimicronucleatum*, syngen 2, and its genetic control. *J. Cell. Physiol.* 67:239.
38. Karakashian, M. W. 1965. The circadian rhythm of sexual reactivity in *Paramecium aurelia*, syngen 3. In *Circadian clocks*, ed. J. Aschoff, 301. Amsterdam: North Holland.

6 Symbionts and Mitochondria of *Paramecium*

6.1 INTRODUCTION

This chapter is devoted to genetic systems in the *P. aurelia* complex based on the presence of symbiotic bacteria in *Paramecium*. Obviously, such mechanisms are very different from the standard genetic system based on DNA-containing genes in the nucleus. As will be shown in this chapter, such systems are common in *Paramecium*. Moreover, they have caused much discussion in the past.

In the 1940s and 1950s (and earlier), the cytoplasm was thought by the overwhelming majority of geneticists to be of little importance in the determination of hereditary characters, and that is still the situation today. On the other hand, many biologists who were not geneticists, especially those studying embryology, at one time held a contrary opinion and believed that fundamental traits of living organisms were controlled by a cytoplasmic part of the genetic system. It was thought by some that the nuclear genes merely acted as modifiers of the relatively minor superficial variations that had been studied many years earlier by Mendel. According to this view, Mendel had been restricting his attention to relatively minor character variations, like those affecting round or wrinkled peas, or similar supposed trivialities. As an example of this view, we may mention a report by S. Wright[1] (p. 301): "It was often asserted a few years ago that the Mendelian genes determine only superficial individual differences, while the fundamental plan of development is cytoplasmic in its heredity." This extreme view was regarded at that time as untenable by nearly all geneticists, including Wright and even Sonneborn.[2,3] See also Beale.[4] To some extent this was a healthy reaction to the rigid mechanistic concepts of the early *Drosophila* geneticists, with their naive notions of genes as "beads on a string," which apparently went periodically through formal rituals involving dividing, breaking, rejoining, and so on, and were apparently uninfluenced by the external environment. The remaining parts of the cell, those outside the chromosomes, seemed, according to this view, to be no more than a kind of tank into which the genically controlled products were discharged. Thus, Morgan[5] wrote, somewhat disdainfully: "The cytoplasm can be ignored genetically." Altenburg and Muller,[6] who were, of course, also members of Morgan's *Drosophila* school, had been somewhat more cautious than Morgan himself and went to great trouble to prove that even the most elusive hereditary characters of *Drosophila*, such as one called truncate, were governed by genes located at precisely identifiable chromosomal sites. Therefore, these workers (Altenburg and Muller) also concluded that heredity of *Drosophila* was controlled primarily by factors in the chromosomes, although Muller himself clearly kept an open mind on the subject.[7]

However, in plants it had been known for many years, from work on chloroplast variations by Correns[8] with *Mirabilis*, and by Baur[9] with *Pelargonium*, that crosses between different strains of a species produced hybrids differing according to which strain was used as male and which as female parent in crosses. There was therefore evidence for the existence of extranuclear genetic determinants of chloroplasts. Whether the sites of these factors were in the cytoplasm at large or inside the chloroplasts themselves was in 1909 a matter of dispute, but this is now considered settled in favor of determinants within the chloroplasts themselves. In *Paramecium*, chloroplasts are found only in the symbiotic algae of *P. bursaria*. The botanical literature also contained accounts of experiments indicating that reciprocal crosses between different strains or wild varieties produced hybrids that differed in characters other than those affecting the chloroplasts. This conflicted with the findings of G. Mendel, who had shown that it was unimportant, in peas at any rate, whether a given hereditary factor was transmitted through the male or female parent. Precise genetic analysis of these aberrant, non-Mendelian phenomena was, however, not carried out at that time (see Correns[10] and Michaelis[11]). As regards animals, practically no cases of cytoplasmic inheritance were known at all, at least in America (see Wright[1]), though Sonneborn[2] held exceptional views on this, and in Germany, as compared with America, the situation was viewed somewhat differently (see Caspari,[12] Oehlkers,[13] Laven,[14] Sapp,[15] and Harwood[16]).

6.2 THE PLASMAGENE HYPOTHESIS

Therefore, the discovery by Sonneborn[17–19] of the "killer" paramecia, which appeared to be controlled by a cytoplasmic genetic factor, or plasmagene, called kappa, aroused much interest. Sonneborn had observed, when carrying out experiments on mating types by mixing together different stocks of *Paramecium* in the same culture vessel, that certain stocks made the medium poisonous to other stocks. Unlike some of the botanists mentioned above, Sonneborn was in a favorable position to make a rigorous genetic study of his killer phenomenon, because of the technical advantages of *Paramecium* for genetic analysis, especially as regards separating and analyzing the roles of cytoplasmic and nuclear genetic factors. Sonneborn found that some of his stocks in *P. biaurelia* and *P. tetraurelia* liberated into the medium a substance that was toxic to other *Paramecium* stocks. He separated his stocks into the two classes killers and sensitives. The sensitives were killed by the killers, but the killers were resistant to killing by cells like themselves. A thorough genetic analysis of the phenomenon was feasible since during conjugation the uptake of substances via the mouth does not take place and sensitives during that time are resistant to the toxin. Crosses between killers and sensitives could therefore be made and did not necessarily result in death of the latter, or their progeny. In this way Sonneborn found clear evidence, which is described below, in favor of a cytoplasmic genetic factor called kappa, controlling the ability of a *Paramecium* to act as a killer. See Figure 6.1.

It was found that when *P. tetraurelia* stock 51 (killer) was crossed with stock 32 (sensitive), the F1s thereby obtained were killers if the F1s' cytoplasm was derived

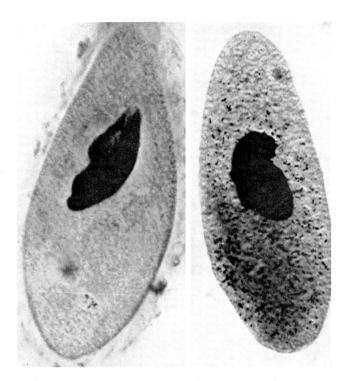

FIGURE 6.1 Light microscope view of a stained sensitive and a killer. Numerous kappa particles are found in the killer. Published with permission of Palgrave Macmillan, from Sonneborn, T. M., *Heredity*, 4, 11, 1982, Figures 1 and 2.

from the stock 51 parent, and sensitives if the cytoplasm of the F1s was derived from the stock 32 parent. By marking the parental cells with India ink, the parental origin of the cytoplasm of the F1s could be easily identified (see Figure 6.2).[20] Usually no cytoplasm passes between two conjugating paramecia, but occasionally a broad cytoplasmic bridge is formed between the conjugants, and a certain amount of cytoplasm can then be seen to be transferred from one conjugant to another. When this happens, both exconjugants and their descendants subsequently obtained by asexual fission are found to exhibit the killer character. Kappa was therefore shown to be inherited through the cytoplasm, and in the view that was held in those days, kappa seemed to be a plasmagene.

When stocks 51 (killers) and 32 (sensitives) were crossed as described above, and autogamy was induced in the offspring of the killer F1s, it was found that two types, killers and sensitives, in a 1:1 ratio were produced in the F2 generation, indicating segregation of a dominant gene. This gene was denoted K, controlling killing ability, but introduction by further crosses of the gene K into cells that were kk homozygotes and had become sensitive due to loss of the cytoplasmic factor kappa did not result in their reversion to killers.

As a result of these and other experiments, Sonneborn concluded that the ability of *Paramecium* to act as a killer was controlled by two factors: (1) an agent situated in the cytoplasm, denoted kappa, and (2) the gene K, situated in the nucleus. These

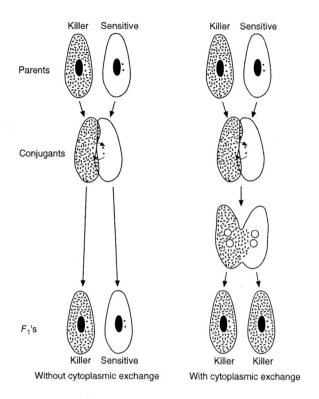

FIGURE 6.2 A cross of killer × sensitive without cytoplasmic exchange (on left) gives no change, while the same cross with cytoplasmic exchange (on right) results in both exconjugants becoming killer. Reprinted with permission of Cambridge University Press from Beale, G. H., *The Genetics of* Paramecium aurelia (London: Cambridge University Press, 1954), 53, Figure 3.

conclusions were confirmed by a series of experiments described in a paper by Sonneborn.[19] The gene K was necessary for maintenance of kappa, but could not cause its production if none was present. Sonneborn therefore cautioned that the killer phenomenon did not justify the assumption of cytoplasmic inheritance in the sense of an independent, self-multiplying determinant. Elsewhere he likened kappa to a "primer in a pump."[21] When a little kappa was put into a *Paramecium* cell, more was formed, and the function of the gene K was considered to control the maintenance and increase of kappa when some was already present, but not to start the production of kappa when none was present previously.

These findings aroused great interest among geneticists at the time, though not universal approval, as shown by the reception of Sonneborn's paper at a symposium at Cold Spring Harbor, Long Island, New York, in 1946.[22] At that time, when the first reports of the killer paramecia were made by Sonneborn, somewhat similar nongenic and cytoplasmically inherited phenomena were being described by others working with other organisms, for example, the "petite" yeasts of Ephrussi et al.[23,24] and the CO_2-sensitive *Drosophila* of L'Héritier,[25,26] as well as the work of Spiegelman et al.[27] on enzyme adaptation in yeast. At this time the plasmagene theory was proposed, which assumed that mRNA (or a DNA copy of the gene) was self-reproducing. The subject was therefore one arousing interest among geneticists, and it was realized that if kappa were indeed a normal part of the genome, then perhaps something other than chromosomal genes were important in genetics.

One criticism of Sonneborn's findings, made especially by Altenburg and Lindegren (see Sonneborn[22] and Altenburg[28]), was that kappa was not a normal part of the genome. Instead, it was held to be a virus or a foreign symbiotic organism rather than a plasmagene. Sonneborn vigorously opposed such suggestions. The fact that kappa is but one example of a "normal and regular" system of cytoplasmic factors in *Paramecium* seems to exclude the virus interpretation.[22] His belief that kappa was part of a normal and regular genetic system was, however, based on other studies that he was making at that time on *Paramecium*, namely, on mating types and antigens, but the supposed similarities in inheritance of kappa with that of these other characters were seen to be superficial, as was soon realized by Sonneborn himself. Thus, he concluded[3] that the killer model does not apply to the antigenic types. He was unwilling to accept the virus or symbiont model for kappa at that time, though later he changed his views on the matter, due to the findings of Preer and colleagues.[29-32]

Kappa particles, especially those occurring in *P. biaurelia*, could grow at varying rates depending on the amount of food in the medium and on the temperature. They were sometimes unable to multiply at the same rate as their host paramecia, so that the kappas were diluted out and finally lost, converting their host from killers to sensitives.[30] X-ray experiments[31] showed that they were apparently similar in size to *Rickettsia* or small bacteria, and examination by light and phase-contrast microscopy showed the presence of some clearly visible microscopic cytoplasmic particles that were present in killers but absent in sensitive (i.e., nonkiller) paramecia. These particles could be stained by the Feulgen method, and therefore contained DNA. See Figure 6.1.

After this discovery, Sonneborn[3] described kappa as being "intermediate between an adventitious, infectious virus and a normally present non-infectious cytoplasmic gene" (p. 167). He believed that kappa was transmitted by the method of a genetic particle, that is, by heredity, not infection. This statement arose from a suggestion of Darlington,[33] who had held that an important criterion for distinguishing between cytoplasmic genetic units and symbionts was their mode of transmission from one host to another: if it was via the germ cells, a particle would, in Darlingon's opinion, be a hereditary unit, or gene, but if via the exterior, by infection, a particle would be a foreign organism. At that time, the nature of these biological units, genes, viruses, and so forth, was less clear than it is now. Sonneborn also added that careful studies by Preer and Stark[33] had revealed marked differences between kappa and various types of organisms (algae, bacteria, *Rickettsia*, viruses, etc.) with which various authors had, as he put it, speculatively, claimed to identify it. He then preferred to classify kappa not as a plasmagene, but as a plasmid, according to the terminology of Lederberg[34] at that time. It was a model factor in the borderland between genetics and parasitology, as Sonneborn then phrased it.[35]

In the end, however, Sonneborn had to accept that kappa was a symbiont. He summarized the various changes in his views on the subject[36] by quoting from his earlier paper "immediately after discovering killers, Sonneborn conceded the possibility that their basis might be a symbiotic organism, but he rejected this possibility when efforts at infection failed. A decade later he succeeded in making infection, but nevertheless held that better evidence was required to validate the symbiont interpretation.[37] The evidence that kappa was indeed a bacterium came largely from the work of Preer and his associates, beginning with the characterization and visualization of kappa, and going on to the discovery of its structure, composition, and properties" (p. 491), as described below. Compelling evidence that kappa was a bacterium was provided by Kung,[38,39] who showed that kappa had the cytochromes and other enzyme systems characteristic of bacteria. Further studies on other endosymbionts made it clear that kappa was one of many kinds of bacterial endosymbionts that occurred in many collections of the *P. aurelia* group.

Sonneborn conceded the symbiotic interpretation of kappa only when the evidence had become overwhelming, and that took a very long time, so far as he was concerned. It is probably difficult for modern readers to appreciate Sonneborn's conceptual difficulties over this matter, but he was at one time very keen to establish the existence of plasmagenes. It should, however, be mentioned that many years previously, Müller[40] had first published evidence for the existence of symbionts in ciliates (of course without any information about their genetics.) Needless to say, acceptance of kappa particles as symbionts rather than as plasmagenes, though admittedly a fundamental point, does not in any way reduce our interest in them. It should be stressed here that plasmagenes and symbionts are now seen to be very different concepts, but this was formerly not so clear.

6.3 THE KINDS OF SYMBIONTS

Many of the endosymbionts of *Paramecium* possess or are capable of producing structures visible in the phase microscope called R bodies.[32,41,42] R bodies are of importance in considering the killer phenomenon and will be considered in some detail in the next section. Many of the symbionts were at first considered to be genetic units and were denoted by Greek symbols, such as kappa, mu, lambda, nu, alpha, and so on, in conformance with an earlier convention according to which cytoplasmic genes were depicted by Greek letters and nuclear genes by Latin ones. Later, because it was thought that the symbionts were actual organisms, they were given binomial names by Preer et al.[43] See Table 6.1 for an up-to-date listing of the symbionts. Table 6.1 is based on Preer et al.,[43] Preer and Preer,[44] Schmidt et al.,[45] Quackenbush,[46] Schmidt et al.,[47] and Kush et al.[48] Thus, stock 51 kappa was named *Caedobacter taeniospiralis* (subsequently changed to *Caedibacter taeniospiralis*) instead of kappa. In this account both the original Greek letters and the Linnaean nomenclatural systems will be used, to facilitate reference to old and new papers. Many different types of symbiont with diverse properties are now known in various members of the *P. aurelia* species group and in other ciliates. Görtz has also pointed out that many of the symbionts of *Paramecium* have a smaller genome size than free-living bacteria, as shown by Soldo and Godoy,[49] suggesting that these associations of symbionts and host are very ancient.[50]

Linnaean names given here for the symbionts of the *P. aurelia* complex of species are those used by Preer and Preer,[44] later revised by Quackenbush,[51] and included

TABLE 6.1
The Symbionts

Name of Symbiont	Greek Letter	P. aurelia complex Species	Stocks	Outer end	Unroll from	Found in	Killing
Caedibacter taeniospiralis	Kappa	4	51, 47, 116, 169, 298, etc.	Sharp	Inside	Cytoplasm	Hump
Caedibacter varicaedens	Kappa	2	7, 562, 1038, 50, 511, etc.	Blunt	Outside	Cytoplasm	Spin, vac, paral
Caedibacter pseudomutans	Kappa	4	51m1	Blunt	Outside	Cytoplasm	Spin
Caedibacter paraconjugatus	Mu	2	570	Blunt	Outside	Cytoplasm	Mate
Caedibacter caryophila	Kappa	9	BGD19 (*P. novaurelia*)	Blunt	Inside	Cytoplasm	Paral
Caedibacter caryophila	Kappa	*P. caudatum*	C231 (*P. caudatum*)	Blunt	Inside	Macronucleus	Paral
Pseudocaedibacter conjugatus	Mu	1, 8	540, 548, 551, 555, 131	Absent		Cytoplasm	Mate
Pseudocaedibacter falsus	Nu, Pi	2, 4, 5	1010, 87, 314, 51, 139, etc.	Absent		Cytoplasm	None
Lyticum flagellatum	Lambda	4, 8	239, 216, 229, 299, 327	Absent		Cytoplasm	Rapid lysis
Lyticum sinuosum	Sigma	2	114	Absent		Cytoplasm	Rapid lysis
Tectibacter vulgaris	Delta	1, 2, 4, 6, 8	561, 1035, 225, 131, etc.	Absent		Cytoplasm	None
Holospora caryophila	Alpha	2	562	Absent		Macronucleus	None

The table presents the symbionts in the species of the *P. aurelia* complex complex plus one found in both *P. caudatum* and the complex.
Data from the following sources:

Preer, Jr., J.R., Preer, L.B., and Jurand, A., *Bacteriol. Rev.*, 38, 113, 1974.

Preer, Jr., J.R. and Preer, L.B., *Int. J. Syst. Bacteriol.*, 32, 140, 1982.

Schmidt, H.J., Görtz, H.-D., and Quackenbush, R.L., *Int. J. Syst. Bacteriol.*, 37, 459, 1987.

Quackenbush, R.L., Endosymbionts of killer paramecia, in *Paramecium*, Görtz, H.-D., Ed.,Springer-Verlag, Berlin, 1988, 406.

Schmidt, H.J. et al., *Exper. Cell Res.*, 174, 49, 1988.

Kush, J. et al., *Microb. Ecol.*, 407, 330, 2000.

in *Bergey's Manual of Systematic Bacteriology*.[52] In this connection, it is important to summarize the bacterial affinities of the endosymbionts of *Paramecium*, as was done by Preer et al.[43]

The relevant data are as follows. The size and shape of the symbionts are typical of bacteria; the symbionts lack mitochondria and nuclear membranes, though most are seen by electron microscopy to be surrounded by one or two unit membranes, which may vary in different forms. All the symbionts tested have been found to be Gram negative, though there may be some variability in their reaction to Gram's stain. They contain ribosomal RNA with sedimentation coefficients typical of bacteria and not of *Paramecium*. Their ribosomal RNAs hybridize with the DNA of the bacterium *E. coli*, but only weakly with the DNA of *Paramecium* itself. Isolated kappa endosymbionts, like bacteria, respire, can utilize glucose and contain most of the enzymes of the glycolytic pathway. The pentose phosphate shunt and the citric acid cycle have been demonstrated in kappa and mu, and have an electron transport system with cytochromes quite different from those of *Paramecium* itself, yet virtually identical to those of certain bacteria. Structures like flagella are found on certain symbionts, such as those called lambda, sigma, and delta. Some of the symbionts have been shown to contain respiratory enzymes typical of bacteria rather than of higher organisms.[39] Some symbionts of *Paramecium* have been shown to be sensitive to antibiotics as reported by Sonneborn.[37] In many characters, therefore, the symbionts of *Paramecium* are typical of bacteria. *Paramecium* endosymbionts have usually been found, however, not to be readily transmissible from one host cell to another by infection, as occurs with many bacteria, though such transmissability has been shown to occur with symbionts under certain unusual circumstances.[53] Exceptionally, as with the type of symbiont denoted alpha (see below), intercellular transmission occurs freely.[54] Claims have been made in only two cases, mu and lambda, that the symbionts of *Paramecium* can be cultured in nutrient medium outside their ciliate hosts.[55,56] Questions about the validity of such claims have been expressed.[43] It has also been shown by Gibson[57] and Koizumi[58] that endosymbionts can be easily transferred from cell to cell by artificial micromanipulation. Thus, in many respects, the endosymbionts of *Paramecium* resemble bacteria.

In one case it has been reported[59] that the symbiont lambda in stock 299 of species *P. octaurelia* apparently supplied its host *Paramecium* with folic acid, since lambda-free paramecia were unable to grow in an axenic medium lacking folic acid, while lambda-containing paramecia were able to grow without such added folic acid. Thus, it is possible that the lambda endosymbionts supply at least some of the nutritional requirements of the host paramecia.

6.3.1 THE FIRST KILLERS FOUND IN *P. BIAURELIA*

In today's nomenclature the symbionts of *P. biaurelia* are called *Caedibacter varicaedens*, but of course when killers were first found, even the fact that symbiosis was involved was unknown. The story began, as noted in Section 6.2, when Sonneborn mixed various strains of the *P. aurelia* group in trying to find mating types. What he found was that sometimes mixtures involving *P. biaurelia* resulted in killing one of the strains in the mixture.[18] The animals that were killed often revealed different and characteristic prelethal effects. These effects were sometimes rapid spinning on their longitudinal axes, in other cases, the formation of large internal vacuoles, or simply an inhibition of swimming rate culminating in paralysis.

Crosses between killers and sensitives were difficult to interpret at this time because of the lack of marker genes in *P. biaurelia* stock 7. However, marker genes were available in *P. tetraurelia* stock 51, where the discovery of kappa and the *K* and *k* genes was first made (see next section). It was not until much later, after marker genes had been found in *P. biaurelia*, that Balsley[60] found the gene *K* in *P. biaurelia* stock 7 killer. When stock 7 killer (genotype KK) was crossed to sensitive stock 1010 (genotype kk) and autogamy was induced, kk segregants (sensitives) and KK (killers) were produced in equal numbers, showing typical Mendelian inheritance.

Even in the early 1940s, however, problems with the notion that kappa was an intrinsic part of *Paramecium* began to appear, for it was found that under conditions of maximum food supply, the *Paramecium* of stock 7 of *P. biaurelia* could grow at about three fissions per day, while the kappas could only multiply at about two fissions per day, leading to a gradual dilution of kappa, and its final complete elimination and conversion of the paramecia to permanent sensitives.[30]

As pointed out above, one of the most striking observations obtained by microscopic examinaton was the existence of inclusions known as R bodies, tiny (about 0.5 μm in size), highly refractile inclusions that were present in up to about 30% of the kappa particles present in a given stock.[32] Kappa particles containing the R bodies were called brights, while kappa particles lacking them were called nonbrights. The R body had an interesting structure, for viewed in one direction it was doughnut shaped, and when it rotated 90° its optical appearance changed to that of two parallel lines. Thus, it was concluded that its true three-dimensional structure was that of a doughnut, and the two parallel lines were the walls of the doughnut when the doughnut was seen edge-on. Observations with the electron microscope[41] fully confirmed this structure and revealed that the R bodies of *P. biaurelia* stock 7 actually consisted of a roll of tape, pointed at the inside end and blunt at the outside end. They often unrolled spontaneously and irreversibly, or could be induced to unroll by an extract of *Paramecium*. When unrolled they formed a loose, open spiral tape. See Figure 6.3. Moreover, the R bodies had numerous adhering phage-like structures on their inner portions that were assumed to be bacteriophages.[41] Some were filled with DNA, and some appeared empty. The particles were isolated by centrifugation, extracted, and their DNA was isolated.[61] Also, high magnification revealed numerous small structures on the surfaces of unrolled R bodies. These were judged to be unassembled units of phage capsids, or perhaps unassembled units of R bodies.[62] See Figure 6.4.

Hybridization studies by Quackenbush[46,63,64] suggest that although there are substantial differences between the different killers, the DNAs of stock 7 (a spinner killer), stock 562 (a lysis killer), and stock 1038 (a paralysis killer) could all be included within one species. Numerous other stocks have proved to be killers in *P. biaurelia*, but except in special cases have all been given the name *Caedibacter varicaedens*. Such killers are very common in *P. biaurelia*, and more than half of the collections of that species from nature prove to have symbionts.[65]

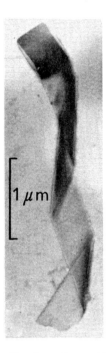

FIGURE 6.3 7 kappa R body unrolling from the outside. Note the blunt end of the outside of the roll of ribbon. Reproduced with permission of the Company of Biologists from Preer, L. B., et al., *J. Cell Sci.*, 11, 593, 1972, Figure 5.

Evidence also made it clear that R bodies play a role in the killing of sensitives. Sensitive paramecia were still killed when exposed to lysates of killers of stock 7, freeing the bright kappas.[66] They found that centrifugation of such lysates sedimented the killer toxin. Careful measurements of the rate at which the toxin was sedimented showed that the toxin followed the bright particles, and not the nonbrights. The toxin was not a soluble substance but was the bright particle itself. Moreover, lysis of bright particles freeing the R bodies revealed that the toxin still was easily sedimented, but now with the R bodies themselves. Further lysis of the R bodies, however, resulted in loss of all killing activity. The role of kappa in killing was suggested by examination of sensitives after they had ingested killer particles. Food vacuoles in sensitives were found with unrolled R bodies that appeared to penetrate the wall of the food vacuole.[67] The following scenario is then suggested for kappa killing: brights are liberated from killers into the medium by way of the cytopyge, taken up by the mouths of sensitives (nonfeeding paramecia are not killed) where R bodies then suddenly unroll, the killer toxins cause breaks in the food vacuole membrane, and the toxin is released to the cytoplasm. Breaks in the food vacuoles of killers are not obtained when they ingest R bodies. Although much of this explanation is admittedly speculative, it is consistent with all the facts. Why each killer is resistant to its own killing is unknown, but remains an interesting question.

As pointed out, bright kappa particles are the killer particles and have no capacity to reproduce. However, nonbrights have clearly been shown to be capable of

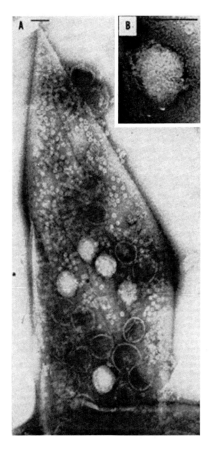

FIGURE 6.4 Stock 7 R body with phage-like bodies and capsomeres. (A) Inner tip of a stock 7 R body with large phage-like bodies and small capsomeres. (B) Inset shows a hexagonal-shaped phage-like body. From Preer, J. R. Jr. and Preer, L.B., *Proc. Natl. Acad. Sci. U.S.A.*, 58:1775, 1967, Figure 1 A and B.

reproduction in *P. tetraurelia*, for when injected into sensitives they convert them to killers.[68] Thus, the kappas act like strains of bacteria containing inducible phages. Presumably after the proper stimulus, the phage DNA in nonbrights is induced and produces R bodies as well as phages, and the individual bacterium dies. See Figure 6.4.

6.3.2 THE KILLERS FOUND IN *P. TETRAURELIA*

As noted in Section 6.3.1, marker genes were available in the killers of stock 51 of *P. tetraurelia*, and this quickly led to the establishment of the cytoplasmic factor, kappa, and the gene *K* and its allele *k*.[19] Now we refer to the symbionts *of P. tetraurelia* as *Caedibacter taeniospiralis*. Killers of *P. tetraurelia*, like the killers of *P. biaurelia*, have R bodies, but the R bodies proved to be quite different. Moreover, the hybridization studies of Quackenbush[64] showed that the kappas of *P. tetraurelia*

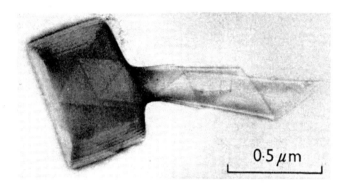

FIGURE 6.5 Stock 51 R body unrolling from the inside. 51 R bodies unroll from the inside, to produce a tube-like structure. Reproduced with permission of the Company of Biologists from Preer, L. B., et al., *J. Cell Sci.*, 11, 591, 1972, Figure 1.

were markedly different from those of *P. biaurelia*. Unrolling of the R bodies of *P. tetraurelia* did not proceed in the same way as unrolling of the R bodies of *P. biaurelia*. See Figure 6.5. They unrolled from the inside like a roll of sticky flypaper, producing a long, tight tubular structure 10 to 15 μm long,[42] with both ends pointed rather than having one end blunt. Moreover, when the pH was lowered they instantly unrolled, and when the pH was brought to alkaline they instantly rerolled.[42] This unrolling and rerolling could be brought about repeatedly.

The kappas of *P. tetraurelia* showed that nonbrights could infect and multiply and that brights could not multiply, but could kill.[68] These facts suggest that the production of brights involves the induction of a temperate phage in the kappa bacterium. The presence of phage-like structures visible in the electron microscope confirms this notion in *P. biaurelia*. However, no sign of phages is found in *P. tetraurelia*. Phages appear to be replaced by covalently closed DNA in the form of plasmids in *P. tetraurelia*. It should be mentioned that the capsid-like structures seen in the *P. biaurelia* kappas are not present in the kappas of *P. tetraurelia*. Instead, helical structures in *P. tetraurelia*,[43] often arranged in parallel rows within the R body, are found. See Figure 6.6. Although it is possible that these structures enclose the plasmid DNA, it appears more likely that they represent unassembled R body proteins or phage capsids.

Under certain conditions, infection of kappa into kappa-free paramecia by nonbright kappas may be brought about through the external medium, though usually this requires a very dense population of particles in the medium surrounding the cells to be infected, as shown by Sonneborn[53] and confirmed by Tallan[69] and Mueller.[68] For such infection to be achieved, certain other special conditions, such as an excess of food in the medium, are required.

The kappas of *P. tetraurelia* differ from those of *P. biaurelia* in being able to multiply as fast as the paramecia in which they are found. Sonneborn[22] was able to carry out dilution experiments by culturing stock 51 at an elevated temperature adjusted to inhibit kappa division more than the multiplication of the *Paramecium*.

Moreover, Dilts[70] found that the 51 kappas contained covalently closed circular DNA molecules 13.7 μm in length, or about 28 megadaltons, obviously plasmids

FIGURE 6.6 Capsomere-like structures from stock 51 R bodies. Coils can be seen on the inside surface of an R body. Reproduced with permission of the Company of Biologists from Preer, L. B., et al., *J. Cell Sci.*, 11, 593, 1972, Figure 3.

within the kappas. These plasmids could be isolated by CsCl centrifugation, cut with restriction enzymes, cloned, and so on, using the tools of molecular biology. Quackenbush[71] mapped the plasmids from different strains of *C. taeniospiralis* from around the world (e.g., stock 51 from Indiana, 116 from Indiana, 169 from Japan, 47 from California, 298 from Panama, and A30 from Australia) and found that all were very similar except for a few transposon sequences that were inserted at various places in the plasmid. Another *P. tetraurelia* stock, 169, gave rise to a mutant, 169-1, that had lost its ability to kill and produce R bodies. Studies revealed that a 7.5 kbp transposon had been inserted within the R body coding region of the plasmid, proving that R bodies are required for killing.[72]

Quackenbush and Burbach[73] made the remarkable discovery that when a portion of the plasmid was cloned and transformed into *E. coli*, the *E. coli* would develop R bodies. Later Kanabrocki et al.,[74] using deletions, were able to identify the R body coding region. The section of the plasmid that they injected contained several open reading frames (ORFs) of about 200 bp each, designated *Reb A*, *Reb B*, and *Reb C*. All of these genes were transcribed in *E. coli* if R bodies were made. If any one of these three were left out, then no R bodies were obtained. However, the R bodies produced by *E. coli* were nontoxic. The DNA sequences of the plasmids were ascertained in numerous plasmids found in different parts of the world, and differences as well as similarities were noted, but they were all quite similar.

The finding that the R body and the toxin are coded by the plasmid, and in some cases by a phage-like DNA, raises questions about the validity of the phylogeny of the kappas, which depends on both bacterial and plasmid characteristics. How can we be sure that the plasmid or phage does not move from one kind of bacterium to another? This does not seem to be the case for the 51 type kappas, based on a comparison of a number of different killers collected from different localities (see above), for in all these cases both the kappas and the kappa plasmids are very similar.[63,64,71] Occurrence and variation of kappa particles in natural populations of *Paramecium* have been studied. It has been shown by Sonneborn[53] and confirmed by Tallan[69] and Mueller[68] that kappa particles, defined as R-body-containing endosymbionts, have been found in only two of the fifteen known species of the *P. aurelia* complex, namely, *P. biaurelia* and *P. tetraurelia*, though kappa particles are found frequently in these two species, especially *P. aurelia*, scattered about the world. All the other species of the *P. aurelia* complex have been carefully screened for the presence of kappa, but no species have been found to contain them.

It should be mentioned that structures like the R bodies of *Paramecium* are known to be present in certain free-living bacteria, for example, *Pseudomonas taeniospiralis* and *P. avenae*.[75-77] Such bacteria, however, do not have the toxic properties of killer paramecia.

6.3.3 THE MATE KILLERS

These symbionts have been found in *P. biaurelia* and *P. octaurelia* and are designated *Pseudocaedibacter conjugatus*. A third mate killer has also been found in *P. biaurelia*, and it will be discussed at the end of this section. Unlike the kappa-containing symbionts, those denoted mu do not kill sensitive paramecia that are swimming in the same culture fluid, but kill only when there is close contact between *mu* particle-containing cells and sensitive cells, as occurs during conjugation. Paramecia that contain mu particles are denoted mate killers.

The first examples of this type to be described were those occurring in *P. octaurelia* stock 138.[78] Siegel found that although conjugation between stock 138 mate killers and sensitive paramecia of *P. octaurelia* resulted in the normal process of reciprocal fertilization, and the exconjugant receiving cytoplasm from the mate killer conjugant regularly yielded viable progeny, the other exconjugant either died without dividing or produced a few daughter cells that soon died. The micronuclei of the sensitive conjugants were damaged or destroyed as a consequence of prolonged contact between the surfaces of the mate killer and sensitive cells, and exchange of nuclei was not required for development of the killing effect. Similar mate-killing phenomena were later found with two other *P. octaurelia* stocks (stocks 130 and 131). When any two of these three mate-killing stocks were paired, there was a kind of pecking order of mate killing between them: 138 > 130 > 131.[79] It was also found that some pairings of stocks 130 or 131 with sensitive paramecia did not always produce outright death, but merely some retardation of the exconjugants receiving cytoplasm from the sensitive parents.[80]

The mate-killing paramecia of *P. octaurelia* were shown to contain bacterium-like symbionts, which were in a general way similar to kappa, except that they did not contain R bodies. Thus, Siegel[78] reported that stock 138 mate killers contained between five hundred and two thousand rod-shaped Giemsa-positive particles of length 0.5 to 1.0 μm. He also found that the mu-bearing stocks contained a dominant gene, *M*, and its allele, *m*, was present in sensitives. The genetic basis of mate killing may therefore be considered similar in a general way to that of kappa.

Mate killing has also been found in three stocks of *P. primaurelia*, 540, 548, and 551.[79] In each of these three stocks the cytoplasm may be seen to contain numerous bacterium-like particles, those of stock 540 varying in length from 2 to 20 μm or more, while those from the two other stocks (548 and 551) are much shorter and more uniform in size. All were Feulgen-positive, rod-shaped structures, with distinct cell walls, and lacked R bodies. Interactions between the three *P. primaurelia* types of mate killer are somewhat different from the above-described *P. octaurelia* mate killers in that the *P. primaurelia* forms are all mutually toxic, so that conjugation between any two may result in death or retardation of all the progeny. Of the three, stock 548 is the most virulent, causing death of a sensitive partner within a few hours of conjugation, while stock 540 usually causes death of its partner only after 12 hours, and stock 551 mate killers only after 1 to 2 days, or occasionally not at all.[79,81]

A single mate killer stock has been found also in *P. biaurelia*.[82] Although it is a mate killer, it is not a kappa killer. Not much is known about this type except that the symbiotic particles present in these mate killers have R bodies. Hence, it is uncertain whether they should be classified as mu or kappa. They have been given the Linnaean-style name of *Caedibacter paraconjugatus*.[51,52] It is not known whether the R bodies have anything to do with mate killing.

6.3.4 THE RAPID LYSIS KILLERS

These forms consist of lambda, *Lyticum flagellatum*,[83,84] and sigma, *Lyticum sinuosum*.[85] These two endosymbionts are mentioned here together because of their property of producing rapid lysis of sensitive cells. Paramecia containing lambda produce lysis of sensitive cells after only 30 minutes of exposure. These symbionts do not contain R bodies, and presumably produce their toxic effects by a different mechanism from that of the R-body-containing kappa particles described above. Lambda and sigma are also remarkable for their large size—lambda is 5 μm and sigma is 15 μm long. See Figure 6.7. They are the largest of the *Paramecium* endosymbionts so far known. These symbionts are also remarkable for their possession of numerous structures resembling the peritrichous flagella of bacteria.[86] The function of these flagellum-like structures is unknown; presumably movement of the symbionts in the cytoplasm of *Paramecium* would be very restricted, owing to the viscosity of the latter. They have never been seen to move in squashes of cells on microscope slides.

Lambda has the shape of straight rods, while sigma is curved or spiral. Both lambda and sigma are surrounded by vacuole-like clear areas in the cytoplasm. Lambda occurs in species *P. tetraurelia* and *P. octaurelia*, while sigma has been found so far only in one stock of *P. biaurelia*, stock 114.

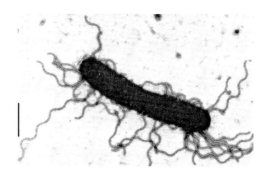

FIGURE 6.7 Stock 327 lambda in *P. octaurelia* showing flagella. The bar on the left is 1 micron. With permission from the American Society for Microbiology; from Preer, J. R. Jr., Preer, L. B., and Jurand, A., *Bacteriol. Rev.*, 38, 21, 1974, Figure 7.

6.3.5 MUTANTS

Workers have found that a stock of killers of one kind sometimes gives rise to killers of another kind. Such occurrences have often been referred to as the production of mutants. It should be remembered, however, that the killers are made up of populations of symbionts, and that none is necessarily pure. Thus, a given killer *Paramecium* may well have more than one kind of symbiont within it, and the production of mutants may only represent the segregation of different kinds of symbiont from a mixed population. It is possible, as suggested by Preer et al.,[43] that the existence of many of the symbiont-containing paramecia is a very ancient phenomenon, and that the populations of symbionts inside many stocks of paramecia have become heterogeneous in the course of time.

Stock 51 was the source of symbionts in addition to *C. taeniospiralis* that were originally mistakenly thought to be mutant (51m1) forms of the kappas described above.[87] These 51m1 symbionts were later shown to be different from the original *C. taeniospiralis* in a number of respects, such as in producing not hump killing, but spinning of affected sensitive paramecia. Moreover, the phage-like particles in *C. pseudomutants* had a spherical shape, like those found in stock 7 killers, rather than helical structures like those found in *C. taeniospiralis*. In view of these differences, 51m1 kappas were reclassified as *C. pseudomutans*,[46] and it is assumed that they arose independently of *C. taeniospiralis*. It may be speculated that at one time stock 51 originally contained a mixture of symbionts now denoted *C. taeniospiralis* and *C. pseudomutans*, and during subculturing the two species segregated into different cultures.

The diverse types of symbionts within a cell can only be separated in the laboratory by a dilution technique, whereby the number of symbionts per *Paramecium* is first reduced to one, by differential multiplication rates of the paramecia and the symbionts. After the level is down to one per cell, food is restricted to the paramecia, allowing the number of symbionts to be restored to the full number in the cells with single particles. However, the method has been used only rarely, as in the separation of 7m1 (a paralysis killer) from its host stock 7 killer (a spinner killer).[88] Whether

FIGURE 6.8 Delta with a single flagellum. The flagellum emanates from the right side of the symbiont and runs forward. The bar on the right is 1 micron. With permission from the American Society for Microbiology; from Preer, J. R. Jr., Preer, L. B., and Jurand, A., *Bacteriol. Rev.*, 38, 124, 1974, Figure 13.

7m1 represents a true mutant or simply a kappa from a heterogeneous mixture of preexisting symbionts is not known.

6.3.6 *TECTOBACTER*: THE SWIMMERS

The most ubiquitous symbiont is found in species 1, 2, 4, 6, and 8. It is known as *Tectobacter vulgaris*. It is easily recognized, for it has what appears in the electron microscope to be a thickened cell wall. However, it is not Gram positive. See Figure 6.8. It also has sparse flagella, which make it mobile in squashes of cells on a microscope slide, and careful observation gives the impression that it is mobile also within the cytoplasm of its host *Paramecium*. What advantages might be imparted by having mobility in the cytoplasm of *Paramecium* is unknown. These attributes are discussed in Preer and Preer.[52] It is not a killer, and is found occasionally along with other symbionts in a given cell.

6.3.7 A NUCLEAR SYMBIONT

An alpha symbiont occurs in stock 562 of *P. biaurelia*.[54] It has been named *Holospora caryophila* and was collected by one of the authors of this book (G. H. B.) in

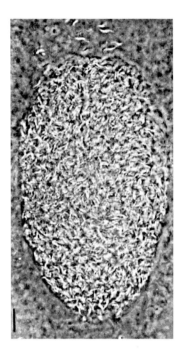

FIGURE 6.9 A macronucleus filled with alpha symbionts. The bar on the left side of the photograph is 10 μm. With permission; from Preer, L. B., *J. Protozool.*, 16, 574, 1969, Figure 10.

a stream near the airport at Milano, Italy, in 1968. It also has been found to occur in other species of *Paramecium*, in both the *P. aurelia* group and *P. caudatum*.[89] The stock 562 alpha symbionts are remarkable for their property of growing in the macronucleus, rather than the cytoplasm of paramecia.[54] See Figure 6.9. These symbionts consist of two different morphological forms: (1) short crescent-shaped structures (or sometimes the crescents are so extreme that they form nearly closed circles), about 2 μm long, and (2) longer twisted or spiral forms reaching 5 μm in length. It is thought that the two forms are interconvertible, and therefore genetically alike. When growth of the host stock 562 of *P. biaurelia* is rapid, only a few—the short forms of the symbiont—are seen, while with slow growth of *P. biaurelia*, many long forms are produced. Alpha particles have no toxic action on other paramecia lacking the symbionts, though the original sample of stock 562 paramecia in which the alpha particles were found consisted nevertheless of killers. This was because stock 562 contained kappa particles in the cytoplasm in addition to the alpha particles in the macronucleus. It is possible to convert paramecia of stock 562 to forms containing only alpha or only kappa particles (or neither), by special treatments, such as growth at a high temperature or with surplus food, and in this way demonstrate that alpha has no killing properties, while the 562 kappa particles (*Caedibacter varicaedens*) do have such properties.

As described elsewhere in this book (Chapter 3), paramecia belonging to any of the *P. aurelia* species groups inevitably pass through autogamy at intervals, and this results in the loss of the existing macronucleus. This must be a hazardous moment in the life of the alpha particles situated in the macronuclei. It has been found, however, that newly developing macronuclei of stock 562, after autogamy, are rapidly colonized by alpha particles from those in the macronuclear fragments of the disintegrating old macronucleus, provided the growth rate of the paramecia after autogamy is slow. Thus, a stock can maintain these symbionts indefinitely. However, if growth after autogamy is rapid, the alpha particles may be lost; presumably in nature the growth rate of paramecia is slow, due to the limited amount of food usually present. The same situation is seen for the *P. biaurelia* killers, such as stock 7, whose kappas can be outgrown by the host *Paramecium*.

It is interesting to note that the infective form of alpha is the long form, for after infection of alpha it is the long form that is always first seen in the anlagen, but shortly afterward the forms within the developing macronucleus are the short ones.

Reinfection of cultures of stock 562 lacking alpha symbionts can be easily brought about by exposure of the paramecia to a homogenate of alpha-bearing paramecia. It has been shown that even after only 1 hour's exposure to such a homogenate, single alpha particles can be seen in the macronucleus of the recipient paramecia. Alpha's capacity to infect suitable host paramecia so speedily is one of the most remarkable properties of these symbionts, distinguishing them from all the others so far known. However, only a limited number of stocks of *P. aurelia* will permit growth of these alpha particles. Of nineteen stocks of *P. biaurelia* tested, only four were found to be capable of being infected by alpha and of maintaining the symbionts. But according to Fokin et al.,[90] some alpha particles, which may not be genetically the same as those found in *P. biaurelia* stock 562, although seemingly identical, can infect paramecia of species 9 as well as *P. caudatum*. The basis of the specificity of the stock 562 alpha particles was shown to be genetic, since a cross between stock 562, which allows growth of alpha, and stock 114, which does not, resulted in a 1:1 segregation of presence or absence of alpha particles in the macronucleus of ex-autogamous F2s, showing that there is a pair of alleles controlling this character.[91]

Alpha particles from stock 562 are specifically adapted to life in the macronucleus of the *P. aurelia* group. They are never found in the micronucleus and rarely in the cytoplasm. Other endosymbionts called *Holospora elegans* and *H. obtusa* are found in the micronuclei and macronuclei of *P. caudatum*.[92] In view of its very rapid passage through the cytoplasm into the macronuclear anlagen during autogamy, alpha has been thought to be motile, though this has not been observed directly, and motility is not said to exist by Görtz and Wiemann.[93] Many years ago, bacterium-like symbionts in the micronucleus of *P. caudatum* apparently similar to alpha, named *Holospora undulata*, were described by Halfkine.[94] Later, apparently (see Wichterman,[95] p. 409), the same species was renamed *Drepanospira Mülleri* by Petschenko.[96]

There have been numerous accounts of additional symbionts in other species of ciliates, particularly *Paramecium caudatum*, but space does not allow us to consider all these cases. See particularly Fokin,[97] Fokin and Ossipov,[98] Skoblo et al.,[99] and Görtz.[100]

6.3.8 A Nuclear Symbiont in One Species and a Cytoplasmic in Another

Caedibacter caryophila was first named as an R-body-containing symbiont present in the macronucleus of *P. caudatum* by Schmidt et al.[45,101] The symbionts make the cells in which it resides into paralysis killers. It was distinguished by the fact that its R bodies were like those of stock 7 of *P. biaurelia* in that they had an inner pointed end and an outer blunt end. However, they unrolled not from the outer end like stock 7, but from the inside like stock 51 kappa (both ends of the R body pointed).

Later, Kusch et al.[102] found what appeared to be the same endosymbiont in the cytoplasm of *P. novaurelia*. This determination was based on the finding that both killers killed the same stocks in the same way. This was true even of a resistant mutant of the form found in *P. caudatum* that was free of R bodies and proved to be resistant and was unaffected by killers from both sources. Sequencing of the 16s rDNA of the form in *P. caudatum* and the form in *P. novaurelia* revealed only one base pair difference. The definitive experiments of putting the symbionts from *P. caudatum* into *P. novaurelia* and vice versa appear not to have been done because of a lack of suitable strains of *Paramecium* known to have the proper maintenance genes.

It is interesting to consider another adaptaton to the symbiotic lifestyle that has occurred in these organisms. Daugherty et al.[103] noted that ATP and ADP, necessary for the energy requirements of most cells, cannot normally be transferred across cell membranes, except in the case of obligate intracellular bacteria, plastids, and mitochondria. Here the genetic determinants may be located not in the bacteria or organelles, but in the nucleus of the host where they produce ATP in the cytoplasm of the host and then transfer it into the symbiont. They noted that in *C. caryophila*, ATP could be transferred across the cell membrane of the symbiont into its cytoplasm, just as in these other cases. Nothing is known about the location of the genes that synthesize ATP in the symbionts. They studied the specificity of the ATP transferring system in *C. caryophila* and found it less specific than in a related symbiont that also has the capacity to transfer ATP across the cell membrane, *Rickettsia prowasecki*, the etiologic agent of typhus.

6.4 THE METAGON

The account of the metagon introduces us to a consideration of an unfortunate and confusing episode in research on the genetics of *Paramecium*. This is the metagon hypothesis. Such incidents are probably more common in scientific research than is generally admitted, but are usually allowed to sink into obscurity. Like the plasmagenes, which were formerly thought to be a useful theoretical basis for kappa, but which are now no longer considered worthy of serious consideration, the metagons were at one time much discussed and, from a historical point of view, have their place in a survey of work on *Paramecium* genetics, though in the course of time they have been discarded, as will be shown below.

This theory starts with work by Chao[104] on the numbers of kappa particles in stock 51 of *P. tetraurelia* in the cytoplasm after replacement of the gene *K* by its allele *k*. He found that there was often a surprisingly long delay between loss of the gene *K*, which was necessary for maintence of the kappa particles, and disappearance of the kappa particles when the gene K was removed. Few ex-autogamous cells had apparently lost their kappa particles before eight fissions after loss of the gene K, and the process was apparently not complete, according to Chao, until after fifteen fissions. It seemed difficult to understand what factors were supporting the kappa particles for such a long period after removal of the gene K. See below for an alternative possibility.

Similar data were subsequently obtained with mu particles in a study of *P. primaurelia* stock 540 by Gibson and Beale.[105] By suitable crosses involving stock 540 (mate killer) and stock 513 (sensitive), mate killer paramecia of genotype $m1/m1$ $M2/m2$ were prepared and backcrossed to $m1/m1$ $m2/m2$ sensitives, yielding a 1:1 ratio for segregation of the $M2$ and $m2$ alleles. Clones of the $m2/m2$ class were first identified by taking samples from each segregating clone and allowing them to pass through sufficient fissions for the mu particles to disappear, while retaining some undivided "sister" (first fission after the cross) cells in nonnutrient medium for later detailed study of the kinetics of particle loss. After establishing that some of these "first fission" cells were of the $m2/m2$ class, researchers subsequently put them into nutrient medium, allowed them to undergo a known number of fissions, and determined the presence or absence of mu particles by microscopic observation of the cells that had undergone these additional fissions. This procedure made it possible to estimate the proportion of cells that still contained mu particles after a known number of fissions following loss of the gene $M2$. Some results of this kind of experiment are shown in Table 6.2. See Gibson and Beale[105] and Reeve.[106]

TABLE 6.2
Loss of Metagons

Number of fissions after loss of gene	Observed percent lacking mu	Calculated percent lacking mu
8	5	1.4
9	17	11.9
10	32	34.4
11	61	58.7
12	71	76.6
13	79	87.5
14	85	93.5
15	93	96.7

The calculated values shown in the table are estimates based on an initial number of 1000 metagons, and computed by Reeve on the assumption that the growth rate of the metagons was 0 after removal of the supporting gene.
From Gibson, I and Beale, G.H., "The mechanism whereby the genes M1 and M2 in *P. aurelia*, stock 540 control growth of the matekiller (mu) particles", *Genet. Res.*, vol. 24, (1962) p. 26, Table I, with permission of Cambridge University Press.

These and other results showed that the rate of loss of mu particles following loss of the gene *M2* was apparently similar to that found by Chao in his experiment on loss of kappa following loss of the gene *K*. There seemed to be no appreciable loss of mu particles before eight fissions after loss of the gene *M2*, and the process was not complete until fifteen fissions had elapsed after removal of the gene *M2*, and its replacement by the recessive allele *m2*. The loss of symbionts, however, when it did occur, appeared to be abrupt. The delay between change of genotype and loss of symbionts was interpreted as being due to presence in the initial cytoplasm of about a thousand entities called metagons, which were considered to be nonreplicating gene products having the determining properties of the corresponding genes.

The results shown in Table 6.2 seemed at first to agree reasonably well with those expected according to the metagon hypothesis, though there were minor discrepancies. However, serious disagreements were found when a more detailed analysis was made with certain of the subclones obtained in the above experiment. For example, the "eleventh fission" cells shown in the above table should, according to the theory, contain only a small number of metagons, and this would result in only a few of the further products of these cells being able to maintain the mu particles after further fissions. The numbers actually obtained, however, were significantly greater than expected. Reeve continued his analysis of the data and was able to show that by assuming a very small growth rate of the metagons, the fit to the data became satisfactory.[106]

On the other hand, attempts to repeat the old work of Chao[104] on the kinetics of loss of kappa particles following removal of gene *K* failed. Another investigator[107] found that in the repeats of the experiments, the kappa particles were usually lost much earlier than had been found by Chao, though admittedly they were occasionally not lost till much later. See also data published long ago by Sonneborn[21,108] on the transfer of cytoplasm during conjugation of cells with and without kappa particles.

In spite of these disagreements, which seemed to be only minor, the metagon hypothesis was at first considered to be acceptable, and further investigations[109,110] were leading to even more remarkable results. These experiments involved feeding a ciliate predator of *Paramecium*, *Didinium nasutum*, with kappa or mu particles and observing the maintenance of these symbionts, and presumably also the metagons, in *Didinium*, even when the latter were subsequently fed on symbiont-free paramecia. It was then assumed that the metagons changed from their previous nonreplicating state in *Paramecium* into one able to replicate in *Didinium*.

It was concluded that the metagon, which had previously been thought to be a nonmutable form of mRNA in *Paramecium*, had become something like a virus in *Didinium*. Sonneborn was very much attracted to this idea, which was a kind of rebirth of his earlier speculations on genes, plasmagenes, and viruses, as expressed in his paper entitled "Beyond the Gene."[3] He went so far as to state:[109] "The metagon appears to replicate like an RNA virus in *Didinium* and to arise as the mRNA of M genes in *Paramecium*" (p. 876). This was reminiscent of his much earlier speculation that kappa was intermediate between an adventitious, infectious virus, and a normally present noninfectious cytoplasmic gene.[3]

In the end, however, the whole idea of the metagon had to be abandoned. Repeats of much of the earler work were done by Beale and McPhail[111] and Byrne,[112] with largely negative results. In particular, efforts to obtain consistent evidence of the kinetics of mu particles in *Paramecium* following loss of the genes *M1* or *M2*, or of kappa particles following loss of the gene *K*, on which the metagon hypothesis was originally based, were largely unsuccessful.

The relation between the maintenance genes and the symbionts represents an unsolved problem in *Paramecium* genetics. The assumption that the genes *K* and *M2* have assumed a positive role in the maintenance of the symbionts was made during a time when plasmagenes were in favor, but some of the observations still seem difficult to explain. Thus, both mu and kappa were retained in paramecia long after loss of their respective supporting genes, sometimes for as many as twenty-seven fissions (Beale, unpublished). The relation between the gene and the symbionts was never spelled out clearly in the plasmagene theory, although at one time it was held that kappa could recombine with its *K* gene in the nucleus.[21]

On the contrary, now that we no longer believe in plasmagenes, we see another possibility that seems just as likely in explaining the relation between the genes and the symbionts. This possibility was first pointed out by Altenburg in his comments on Sonneborn's Cold Spring Harbor paper.[22] It may be that the action of *K* and *M2* is to contribute essentially nothing, and that it is the alleles *k* and *m2* that produce a substance that makes the cytoplasmic environment less favorable for symbiont multiplication. We would have to assume that dominance is incomplete rather than complete, but that is not a problem. Then after these genes are introduced, the inhibitors build up with varying response of the symbionts to these unfavorable conditions and result in their gradual loss. Sonneborn either ignored or misunderstood this comment of Altenburg, and other workers apparently did likewise. Nevertheless, specificity must be postulated on either hypothesis; for example, Sonneborn[2] reported that kappas from *P. primaurelia* could not be infected into sensitives of *P. biaurelia*.

As for the possibility that the metagon might be converted into a replicating form when transplanted from *Paramecium* to *Didinium*, any experiments bearing on the stability or replicability of the metagon could, in the absence of proof of its very existence, have no meaning. On the basis of Altenburg's notion that symbionts were eliminated by the absence of the supposedly dominant allele, there would be no reason to invoke the metagon, for *Didinium* would not be expected to have genes that would specifically eliminate the symbionts, and the symbionts would multiply without involving metagons.

The moral to this story, for which the first author of this book must accept some responsibility, seems to be that when an original theory is presented, it is important first to check and double-check the supporting evidence. It is all too easy to obtain provisional data that seem to fit in with such a theory at first sight. There may have been subconscious selection of results that seemed at first sight to be supportive, but which were in fact only accidentally so. Deliberate selection of data by an inexperienced assistant is, of course, also a possibility. As Fisher[113] has alleged in his analysis of a completely different case, this may have happened even to Mendel, whose numerical results seemed to have agreed too well with expectations according to his theory, though in the same article Fisher also remarks, perhaps rather unkindly, "fictitious data can seldom survive a careful scrutiny." But the great strength of Mendel's theory was that it was basically sound. As regards the metagon, we are left with the fascinating but unexplained phenomenon of the symbiosis of bacterium-like organisms living harmoniously inside the cells of ciliate protists like *Paramecium*, but no detailed account of the mutual interrelations between the two components of this symbiosis (nuclear genes of *Paramecium* and the kappa or mu symbionts). Sometimes the cytoplasm (or nucleus) of the host cell is packed with immense numbers of symbionts, without causing any apparent harm to their host. The molecular basis of this phenomenon is at present completely unknown, and would surely repay further study. Finally, it is perhaps worth pointing out that the metagon hypothesis was first put forward in 1962 when evidence for the existence of messenger RNA had only just been published, and when the molecular details of the theoretical basis of gene expression were first beginning to be unraveled.[114]

6.5 MITOCHONDRIA

6.5.1 INTRODUCTION

It is reasonable to assume that the genetic determinants of kappa, mu, and other endosymbionts of *Paramecium* lie in the DNA of the symbionts themselves, rather than in the nuclei of the host *Paramecium*, though there is at present little evidence in favor of this statement. As described previously (above), genetic variants of some of the symbionts are known to occur, but the genetic basis of these variants has not been analyzed. All that can be said is that certain *Paramecium* genes, such as K (rather than its allele k) in *P. tetraaurelia*, M (rather than its allele m) in species *P. octaurelia*, and other genes in other species, are necessary for the symbionts to be maintained, but this is an all or none effect. No genetic data bearing on the location of the symbiont genes exist. Of course, if the symbionts retain most of their bacterial qualities, then it is likely that the genes for these qualities will be found in their own genomes. However, nothing is known about the detailed heredity of the symbionts. There are practically no molecular data bearing on this matter: the kappa genome has only had the first step toward investigation by sequencing its DNA, along the lines now pursued with great success in the case of numerous organisms, including *Paramecium*. It is to be hoped that in the future this may be done for the symbionts of *Paramecium*, but so far this is largely unexplored territory.

There is, however, another system involving nucleocytoplasmic interactions in *Paramecium* and having some similarities with the endosymbiont system described in this chapter. This concerns the mitochondria. It involves the interactions between the genomes of the mitochondria and the *Paramecium* nuclei. It is now well known from studies with many other organisms, particularly yeast, that mitochondria contain a relatively small but essential section of DNA located within the mitochondria themselves, as shown by Nass and Nass.[115] In recent years our knowledge of the mitochondrial genome has increased greatly, but is based on studies of organisms other than *Paramecium*. However, in this subject, as in so many others, *Paramecium* has its own idiosyncracies.

In the mitochondria of all organisms so far studied, the role of the nuclear genome predominates over that of the mitochondrial genome, even though the mitochondria are certainly essential for the continuation of life of practically all eukaryotic organisms. In the endosymbionts of *Paramecium*, as mentioned above, the situation seems

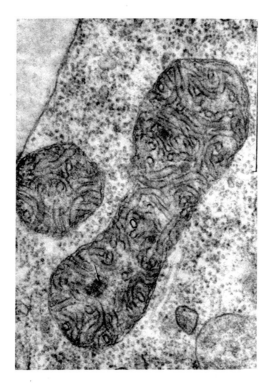

FIGURE 6.10 Two sections through mitochondria from *Paramecium* showing tubular structures on the inside. The mitochondria are about one half micron in diameter and several microns long. When lysed in the external medium, they quickly become irreversibly spherical. Reproduced with permission of Palgrave Macmillan from Jurand, A., and Selman, G. G., *The Anatomy of* Paramecium aurelia (Macmillan, 1969), p. 139, plate 28, Figure 1.

to exist that nearly all their hereditary elements are probably found in the symbionts themselves. In spite of this large difference between the mitochondrial and endosymbiotic systems of *Paramecium*, they can both be studied in the same organism, because both are present in the cells of *Paramecium*. This is particularly relevant in view of the present widespread belief that mitochondria have been thought to have originated by an evolutionary process involving the establishment, at some remote time, of some kind of symbiotic interaction between two previously separate cellular units, and the present mitochondria of many organisms are believed to have arisen originally as endosymbionts.[116] Kappa and other symbionts of *Paramecium* are presumed to have been at one time free-living organisms, though the time at which they first became endosymbionts was probably more recent than that remote epoch when mitochondria are thought to have taken up their residence as organelles in eukaryotic cells. All this is, of course, highly speculative.

In some respects, *Paramecium* is a very suitable organism with which to study mitochondrial genetics. It contains within its cytoplasm some five thousand well-developed mitochondria, which are illustrated in Figure 6.10, and as already discussed, the general biology and life cycle of *Paramecium* makes it particularly suitable for analyzing the separate roles of nuclear and extranuclear genetic factors.

Paramecium has certain disadvantages as compared with yeast, with which most of the original studies on mitochondrial inheritance have been made.[24] But *Paramecium* is an aerobic organism, and hence it can have no variants like the petite yeasts for study of defects in the respiratory system, which constitutes a very important part of the biological function of mitochondria.

6.5.2 Genetics of Mitochondrial Characters in *Paramecium*

The first variant of *Paramecium* that was shown to be controlled by a mitochondrial gene involved resistance to the drug erythromycin. Antibiotics of this type are known to inhibit protein synthesis on ribosomes in the mitochondria, but not on the ribosomes situated free in the cytoplasm (cytosolic ribosomes), which are governed by nuclear genes.

If paramecia are placed in nutrient medium containing 0.25 mg/ml erythromycin, multiplication of the ciliates soon stops, even though they may continue to survive for as long as a month in the erythromcyin-containing media, getting progressively paler and smaller, and moving more and more slowly. Eventually most of the cells so placed die, but after 2–4 weeks in erythromycin, a few paramecia may suddenly resume normal growth, and are then found to be quite unaffected by erythromycin at a concentration of 0.25 mg/ml. The ciliates grow at the same rate as ordinary paramecia do in nutrient medium lacking erythromycin altogether. The cells adapted to growth in medium containing added erythromycin can be shown to be mitochondrial mutants. When they are made to conjugate with normal, erythromycin-sensitive cells, the progeny show cytoplasmic inheritance; ex-conjugants deriving cytoplasm from erythromycin-resistant cells are again resistant to erythromycin, while those deriving cytoplasm from sensitives are sensitive to erythromycin. However, when there is appreciable cytoplasmic exchange between resistant and sensitive conjugants, as may be artificially induced under certain circumstances, both ex-conjugant clones become resistant to erythromycin.[117] The nuclear genome plays no part in the determination of erythromycin resistance, but some genetic factors in the cytoplasm do play such a role. Erythromycin resistance is inherited cytoplasmically, just like the killer character of *Paramecium*, but maintenance of kappa also requires the presence of the gene K, while no comparable gene has been found necessary for the maintenance of a particular kind of mitochondria.

Unfortunately, relatively few characters due to mitochondria other than resistance to erythromycin or chloramphenicol have been identified in *Paramecium*. The organism does, however, have one compensating advantage over yeast for the study of mitochondrial genetics, namely, the suitability of *Paramecium* for the transfer of mitochondria from cell to cell by microinjection, due to the relatively enormous size of the *Paramecium* cells compared to those of yeast and most other organisms.

The technique of microinjection of materials into *Paramecium* was first developed by Koizumi[58] and then modified by Knowles.[118] By this technique it is possible to transfer cytoplasm from cell to cell, not only between paramecia belonging to the same species within which conjugation can take place, but even between cells belonging to different noninterbreeding species.

The details of the inheritance of the two kinds of character—the killers and the mitochondrial variants—are quite different, though both are cytoplasmically based.

In the case of the killer particles, the inheritance is due, as shown previously, to bacterium-like cytoplasmic kappa particles or endosymbionts, which can be seen by microscopic observation, while erythromycin resistance is due to a component of the cytoplasm that cannot be readily identified. Cytoplasmic inheritance of mitochondrial characters was easily proved, not only by the results of crossing experiments with and without cytoplasmic exchange, but also by injecting partially purified mitochondria derived from erythromycin-resistant paramecia into erythromycin-sensitive cells, and selecting the recipients in medium containing erythromycin. After a short time, the injected cells become resistant to erythromycin.[119]

Later experiments proved that the determining element in the cytoplasm of erythromycin-resistant paramecia was mitochondrial DNA. This finding was made by studying the effect of transfer of erythromycin-resistant mitochondria derived from different species of the *P. aurelia* complex into sensitive cells, whose mitochondrial DNA differed from that of the donors in certain restriction enzyme patterns.[120] It was then found that the transformed recipient paramecia came to contain mitochondrial DNA that was like that of the donor cells and different from that of the previously sensitive recipients.

Apart from erythromycin, resistance to two other antibiotics, chloramphenicol and mikamycin, is inherited in the same way as resistance to erythromycin.[117,121] Furthermore, double resistance of mutants to chloroamphenicol (Cr) and erythromycin (Er) was produced by crossing ciliates resistant to Cr and ciliates resistant to Er, and allowing cytoplasmic exchange. These double mutants were resistant to each antibiotic separately, but when exposed to both simultaneously there was no complementation and the cells were killed as shown by Beale et al.,[122] Adoutte and Beisson,[123] Beale,[121] and Adoutte.[124]

On the other hand, if a group of cells resistant to one antibiotic were grown in the presence of the other antibiotic, resistance to both antibiotics could be obtained. In this case, the cells had only one kind of mitochondrion. It carried the genes for resistance to both antibiotics, and unlike the previous case, the cells survived when exposed to both antibiotics.

Such doubly resistant mitochondria, those resistant to a mixture of the two antibiotics, were not produced by recombination between mitochondrial genes in *Paramecium*, as occurs in yeast, where the mitochondria may fuse together at certain stages. In *Paramecium* mitochondria remain separate at all times, even when erythromycin-resistant and chloromycetin-resistant cytoplasms are mixed together in one cell. Moreover, there is no complementation between the two different kinds of mitochondria in mixtures. This has been thoroughly studied and confirmed.[121,123,124] Due to the absence of recombination involving different mitochondrial genes of *Paramecium*, mapping of the mitochondrial genome by classical recombinational methods has not been possible, as contrasted with yeast, where recombination between mitochondrial genes occurs frequently.

A fair number of mitochondrial enzymes are known to be under the control of the mitochondrial, rather than the nuclear, genome. In *Paramecium*, it is difficult to obtain evidence of this by classical genetic means, on account of the paucity of mitochondrial variants within a given species. In spite of the enormous populations of paramecia extending over practically the whole world, and the collection of many hundreds of different stocks, very few spontaneous variants affecting mitochondrial characters have been found in wild populations. This contrasts markedly with the large number of variants affecting some other characters of *Paramecium*, such as the immobilization antigens (see Chapter 7). Different mitochondrial enzymes in a large number of stocks and species of the *P. aurelia* complex were compared by Tait,[125-127] who found that there was little variation within single species, but more between different species. Genetic analysis by classical Mendelian methods is, of course, usually only possible when there is intraspecies variation.

In the few cases where the mitochondrial enzymes of different electrophoretic forms have been found to exist in different stocks of a single species, genetic analysis has been done. Tait[125-127] showed that three mitochondrial enzymes—(1) HBD (beta-hydroxy-butyrate-dehydrogenase), (2) ICDH (isocitrate dehydrogenase), and (3) GDH (glutamate dehydrogenase)—all have variants in different wild stocks of single species. However, variants of this type are quite rare, and were only found after screening about a hundred different stocks of *P. primaurelia* and *P. tetraurelia*, and smaller numbers of thirteen other species. Detailed genetic studies were made of the enzymes HBD, ICDH, and GDH, and all the variants of these enzymes that were studied were shown to be controlled by the nuclear rather than the mitochondrial genome.

One example of an interspecies variation affecting enzyme characteristics is the mitochondrial enzyme fumarase, which differs electrophoretically in *P. primaurelia* and *P. septaurelia*, as shown by Knowles and Tait.[122] Since crosses between the two species do not yield viable progeny, the fumarase variation could not be analyzed genetically by crossing, but had to be studied by exploiting the technique of microinjection of mitochondrial fractions into different cells of these two species. It was then shown that there was no correlation between the type of fumarase variant present and the mitochondrial genomes, but there was such a correlation with the nuclear genomes. Such demonstrations are only feasible when the mitochondria from one species are compatible with other components of another species, as fortunately happens with these two species (see below).

Studies have also been made on resistance to a tetrazolium salt (TTC), which was correlated with various defects in mitochondrial functions, such as cytochrome content and cyanide-insensitive respiration, as well as some interactions with susceptibility to erythromycin and chloramphenicol. Four mutants affecting these TTC characters have been analyzed by Ruiz and Adoutte,[128] and all were found to be due to nuclear, not mitochondrial, genes.

All the mitochondrial enzymes showing electrophoretic variations so far analyzed in *Paramecium* have been found to be controlled by the nuclear rather than the mitochondrial genome, though the data on this are rather scanty. Sometimes, however, there are complex interactions between nuclear and mitochondrial genetic factors affecting the same character. This was apparent from some experiments on the survival of mitochondria from one species of the *P. aurelia* complex after injection into cells of another species. For example, mitochondria from *P. primaurelia* and *P. pentaurelia* were successfully transferred between cells of these two species and *P. septaurelia*, but mitochondria of species *P. septaurelia* failed to develop in *P. primaurelia* or *P. pentaurelia*.[129] Such experiments indicated that the compatibility of mitochondria from different species was mainly under control of the nuclear genome, with some minor effect of the mitochondrial genome. No success was obtained in any transfer experiments involving mitochondria of *P. tetraurelia* into other species.

Further information on the interaction between the nuclear and mitochondrial genomes in the control of mitochondrial characters has been provided by Sainsard et al.,[130] Sainsard,[131] Sainsard-Chanet,[132,133] and Sainsard-Chanet and Knowles[134] on a slow-growing (*cl-1*, or croissance lente) mutant of *P. primaurelia*. *cl-1* is a nuclear mutation showing severe effects on mitochondria. The changes determined by *cl-1* involved a morphological abnormality of the mitochondrial cristae, a deficiency in cytochrome oxidase, and a marked reduction in growth rate. The *cl-1/cl-1* stock now were shown to undergo spontaneous mutations in their mitochondria. These mutations, which were cytoplasmically inherited, occurred at a low rate and were usually different from each other (M^{cl}, M^{su}, M^*). It was concluded that the nuclear and mitochondrial genomes exercised a mutual collaboration in the control of some mitochondrial characteristics.

Tait[135] and Beale et al.[119] found that erythromycin resistance was due to a modification in a mitochondrial ribosomal protein gene. This finding was subsequently confirmed by Cummings et al.[136,137] Moreover, Tait et al.[138,139] studied interspecific hybrids, obtained by injecting mitochondria from *P. tetraurelia* into cells of *P. septaurelia*. It was found that mitochondrial ribosomal proteins differed and were controlled by both nuclear and mitochondrial genes.

In general, it may be concluded that while there is some limited control of mitochondrial genes in carrying out the functions of mitochondria, the nuclear genes play a much more important function in this process.

6.5.3 THE MITOCHONDRIAL GENOME IN *PARAMECIUM*

The mitochondrial DNA of *Paramecium* is large, about 40 kb.[140] While most mitochondrial DNAs are circular, the mitochondrial DNA of several genera of protists have been found to be linear. These include *Paramecium* and *Tetrahymena*.[141] *Paramecium* mitochondrial DNA is also unusual because it is replicated by means of a lariat-shaped intermediate, unlike the circular DNA of most mitochondrial DNAs. As in other organisms, most of the components of mitochondria are synthesized under the control of the nucleus of the cell and imported into the mitochondria, more than 90% according to Gray et al.[142] The mitochondrial genome of stock 51 of *P. tetraurelia* has been completely sequenced.[140]

Those functions retained by the mitochondria are concerned either with oxidative functions of the cell or with the protein synthesizing system that is specific to the mitochondria. The genes concerned with oxidation found in the mitochondria of *Paramecium*[141] are cytochrome oxidase genes (*cox 1*, *cox 2*), two of three of the total number of cytochrome oxidase genes found in protist mitochondria; the cytochrome b gene (*cob*); eight of eleven genes coding for NADH reductase genes (*nad*); and one of five genes coding for ATP synthetase genes (*atp*).

Genes concerned with the special system of protein synthesis shown by the mitochondria[141] are for the small and large subunits of the mtRNA. These are both present in *Paramecium* and are present in virtually all mitochondria. They are unusual in being split in *Paramecium*. Ribosomal protein genes are also present (small subunit), four of twelve; ribosomal proteins (large subunit), three of fifteen; and tRNA genes, four of thirty-one, an unusually low number. For example, yeast has twenty-five of thirty-one tRNA genes. It might also be noted that *Paramecium* has seventeen ORFs that have not yet been assigned, but presumably most of these will code for ribosomal proteins.

A peculiarity of *Paramecium* mitochondrial genomes is the absence of any introns, which occur in other mitochondrial genomes; for example, there is one in the middle of the large rRNA gene in yeast, and altogether some thirteen introns in its mitochondrial DNA.[143] The *Paramecium* mitochondria are also unusual in the structure of their gene products. When comparisons are made between the amino acid sequences of predicted *Paramecium* mitochondrial gene products and homoloogous proteins of a wide range of other organisms, the *Paramecium* gene products are seen to be markedly divergent.[144]

6.5.4 THE MITOCHONDRIAL GENETIC CODE

As is discussed elsewhere in this book (Section 7.4.4), the nuclear genome of *P. tetraurelia*, like that of some other ciliates, such as *Tetrahymena* and *Stylonychia*, is unusual in regard to its genetic code. In the universal genetic code, UGA, UAA, and UAG serve as stop codons. In the macronucleus of *Paramecium*, UGA is the stop codon, and UAA and UAG code for glutamine. In the mitochondria of *Paramecium*, the triplet UGA is not the stop codon, but instead codes for arginine. This is true for mitochondria of all species, and is noteworthy for that reason. The ciliates are also unusual in that all their other mitochondrial codons are identical to the universal code. *Paramecium* is unusual in having mitochondria with only this one exception to the universal code, for various other codon differences appear among mitochondria from different organisms.

One can conclude this section by saying that while the *Paramecium* mitochondrial code is different from that of the mitochondrial genomes of other groups of organisms (and the same is true for *Tetrahymena*), these differences are no greater than those occurring among some other organisms. These variations have been reviewed by Fox,[145] Osawa et al.,[146] and Tourancheau et al.[147]

6.6 CONCLUSIONS

How should we now view the bacterial symbionts of *Paramecium*, in light of their long history of controversy and the often unacceptable theories about them? While the bacterial symbionts of the protozoa and mitochondria are considered together here in the same chapter, it should be pointed out that although they are alike in both having been descended from free-living forms, the mitochondria are very different, having been integrated into the cytoplasm at the very base of the phylogenetic tree. This has resulted in most of the genes of mitochondria having been transferred to the host genome, while in the symbionts of *Paramecium* there is no evidence that a single gene has been transferred. It is true, however, that most or all of the symbionts cannot be cultured outside of *Paramecium*. So poorly have these symbionts been integrated that they well may be considered bacteria, and therefore merely wait upon an ingenious investigator to culture them outside of the cell and prove their independence. Two points might be made, however, that argue for a more intimate role between the symbionts and their protozoan hosts. The first is the fact that the presence of kappa in a strain makes the strain resistant to the specific kind of toxin produced by the kappa. Moreover, as we have seen, this fact has implications for the ecology of the killers. Second is the relation between the kappa particles and their maintenance genes, for we still do not know what these genes do, in spite of failed hypotheses (metagons and plasmagenes) to explain the relationship. We are not even sure whether the gene in killers supports the symbionts or the gene in sensitives is unfavorable to the growth of the symbionts.

REFERENCES

1. Wright, S. 1945. Genes as physiological agents—General considerations. *Am. Nat.* 79:289.
2. Sonneborn, T. M. 1950. The cytoplasm in heredity. *Heredity* 4:11.
3. Sonneborn, T. M. 1950. Beyond the gene—Two years later. In *Science in progress*, ed. G. A. Baitsell, 167. 7th ser. New Haven, CT: Yale University Press.
4. Beale, G. H. 1982. Tracy Morton Sonneborn. *Biog. Mems. Fell. R. Soc.* 28:537.
5. Morgan, T. H. 1926. Genetics and the physiology of development. *Am. Nat.* 60:489.
6. Altenburg, E. M. and Muller, H. J. 1920. The genetic basis of truncate wing, an inconstant and modifiable character. *Genetics* 5:1.
7. Preer Jr., J. R. 2006. Sonneborn and the cytoplasm. *Genetics* 172:1373.
8. Correns, C. 1909. Vererbungsversuche mit blass (gelb) grünen und bundblatrigen Sippen bei *Mirabilis, Urtica*, und *Lunaria*. *Z. indukt. Abstamm.-u VererbLehre* 1:291.
9. Baur, E. 1909. Das Wesen und die Erblichkeitsverhältnisse der "Varietates albomarginate hort." von *Pelargonium zonale*. *Z. indukt. Abstamm.-u VererbLehre* 1:330.
10. Correns, C. 1937. Nichtmendelnde Vererbung. In *Handbuch der Vererbungswissenschaft*, ed. E. Baur and M. Hartmann. Leipzig: Verlag von Gebrüder Bornträger.
11. Michaelis, P. 1951. Interactions between genes and cytoplasm in *Epilobium*. *Cold Spring Harbor Symp. Quant. Biol.* 16:121.
12. Caspari, E. 1948. Cytoplasmic inheritance. *Adv. Genet.* 2:2.
13. Oehlkers, F. 1952. Neue Überlegungen zum Problem der Ausserkaryotischen Vererbungen. *Z. indukt. Abstamm.-u VererbLehre* 84:213.
14. Laven, H. 1959. Speciation by cytoplasmic isolation in the *Culex pipiens* complex. *Cold Spring Harbor Symp. Quant. Biol.* 24:166.
15. Sapp, J. 1987. *Beyond the gene: Cytoplasmic inheritance and the struggle for authority in genetics.* New York: Oxford University Press.
16. Harwood, J. 1984. The reception of Morgan's chromosome theory in Germany: Interwar debate over cytoplasmic inheritance, *Medizinhistorischen Journal* 19:3.
17. Sonneborn, T. M. 1938. Mating types in *P. aurelia*, diverse conditions for mating in different stocks; occurrence, number and interrelations of the types. *Proc. Am. Phil. Soc.* 79:411.
18. Sonneborn, T. M. 1939. *Paramecium aurelia*: Mating types and groups; lethal interactions, determination and inheritance. *Am. Nat.* 73:390.
19. Sonneborn, T. M. 1943. Gene and cytoplasm. I. The determination and inheritance of the killer character in variety 4 of *P. aurelia*. II. The bearing of determination and inheritance of characters in *P. aurelia* on problems of cytoplasmic inheritance, pneumococcus transformations, mutations and development. *Proc. Natl. Acad. Sci. U.S.A.* 29:329.
20. Beale, G. H. 1954. *The Genetics of* Paramecium aurelia. London: Cambridge University Press.
21. Sonneborn, T. M. 1945. The dependence of the physiological action of a gene on a primer and the relation of primer to gene. *Am. Nat.* 79:318.
22. Sonneborn, T. M. 1946. Experimental control of the concentration of cytoplasmic genetic factors in *Paramecium*. *Cold Spring Harbor Symp. Quant. Biol.* 11:236.
23. Ephrussi, B., Hottinguer, H., and Chemenes, A. M. 1949. Action de l'acriflavine sur des levures. I. La mutation "petite colonie." *Ann. Inst. Pasteur* (Paris) 76:351.

24. Ephrussi, B., Hottinguer, H., and Tavlitzki, J. 1949. Action de l'acriflavine sur des levures. II. Etude génétique du mutant "petite colonie." *Ann. Inst. Pasteur* (Paris) 76:419.
25. L'Héritier, P. 1948. Sensitivity to CO_2 in *Drosophila*, a review. *Heredity* 2:325.
26. L'Héritier, P. 1957. The hereditary virus of *Drosophila*. *Adv. Virus Res.* 5:195.
27. Spiegelman, S., Lindegren, C., and Lindegren, G. 1945. Maintenance and increase of a genetic character by a substrate-cytoplasmic interaction in the absence of the specific gene. *Proc. Natl. Acad. Sci. U.S.A.* 31:95.
28. Altenburg, E. 1946. The symbiont theory in explanation of the apparent cytoplasmic inheritance in *Paramecium*. *Am. Nat.* 80:661.
29. Preer, J. R. Jr., 1946. Some properties of a genetic cytoplasmic factor in *Paramecium*. *Proc. Natl. Acad. Sci. U.S.A.* 32:247.
30. Preer, J. R. Jr., 1948. The killer cytoplasmic factor kappa. Its rate of reproduction, the number of particles per cell, and its size. *Am. Nat.* 62:35.
31. Preer, J. R. Jr., 1950. Microscopically visible bodies in the cytoplasm of the 'killer' strains of *P. aurelia*. *Genetics* 35:344.
32. Preer, J. R. Jr. and Stark, P. 1953. Cytological observations on the cytoplasmic factor "kappa" in *P. aurelia*. *Exp. Cell Res.* 5:478.
33. Darlington, C. D. 1955. Review of Beale, 1954. *Heredity* 9:284.
34. Lederberg, J. 1952. Cell genetics and hereditary symbiosis. *Physiol. Rev.* 32:403.
35. Sonneborn, T. M. 1955. Heredity, development and evolution in *Paramecium*. *Nature* 175:110.
36. Sonneborn, T. M. 1975. *Parmecium aurelia*. In *Handbook of genetics*, ed. R. C. King, chap. 20. Vol. 2. New York: Plenum Press.
37. Sonneborn, T. M. 1959. Kappa and related particles in *Paramecium*. *Adv. Virus Res.* 6:229.
38. Kung, C. 1966. Aerobic respiration of kappa particles in *Paramecium aurelia* stock 51. *J. Protozool.* 13(Suppl.):12.
39. Kung, C. 1970. The electron transport system of kappa particles from *Paramecium aurelia* stock 51. *J. Gen. Microbiol.* 61:371.
40. Müller, J. 1856. Beobachtungen an Infusorien. *Monatsberichte der Berliner Akademie* 389.
41. Anderson, T. F., et al. 1964. Studies on killing particles from *Paramecium*: The structure of refractile bodies from kappa particles. *J. Microsc.* (Paris) 3:395.
42. Preer, J. R. Jr., Hufnagel, L., and Preer, L. B. 1966. Structure and behavior of "R" bodies from killer paramecia. *J. Ultrastruct. Res.* 15:131.
43. Preer, J. R. Jr., Preer, L. B., and Jurand, A. 1974. Kappa and other endosymbionts in *Paramecium aurelia*. *Bacteriol. Rev.* 38:113.
44. Preer, J. R. Jr. and Preer, L. B. 1982. Revival of names of protozoan endosymbionts and proposal of *Holospora caryophila* nom. nov. *Int. J. Syst. Bacteriol.* 32:140.
45. Schmidt, H. J., Görtz, H.-D., and Quackenbush, R. L. 1987. *Caedibacter caryophila* sp. nov., a killer symbiont inhabiting the nucleus of *Paramecium caudatum*. *Int. J. Syst. Bacteriol.* 37:459.
46. Quackenbush, R. L. 1988. Endosymbionts of killer paramecia. In *Paramecium*, ed. H.-D. Görtz, 406. Berlin: Springer-Verlag.
47. Schmidt, H. J., et al. 1988. Characterization of *Caedibacter* endonucleobionts from the macronucleus of *Paramecium caudatum*, and the identification of a mutant with blocked R-body synthesis. *Exp. Cell Res.* 174:49.
48. Kush, J., et al. 2000. The toxic symbiont *Caedibacter caryophila* in the cytoplasm of *Paramecium novaurelia*. *Microb. Ecol.* 407:330.
49. Soldo, A. T. and Godoy, G. A. 1972. The kinetic complexity of *Paramecium* macronuclear deoxyribonucleic acid. *J. Protozool.* 19:673.

50. Preer, J. R. Jr., 1977. The killer system in *Paramecium*—Kappa and its viruses. In *Microbiology—1977*, ed. D. Schlessinger, 576. Washington, D.C.: American Society for Microbiology.
51. Quackenbush, R. L. 1982. Validation of the publication of new names and new combinations previously published outside the I.J.S.B. *Int. J. Syst. Bacteriol.* 32:266.
52. Preer, J. R. Jr. and Preer, L. B. 1984. Endosymbionts of protozoa, In *Bergey's manual of systematic bacteriology*, ed. N. R. Krieg. Baltimore: Williams and Wilkins, 795. Vol. 1.
53. Sonneborn, T. M. 1948. Symposium on plasmagenes, genes, and characters of *Paramecium aurelia. Am. Nat.* 82:26.
54. Preer, L. B. 1969. Alpha, an infectious macronuclear symbiont of *Paramecium aurelia. J. Protozool.* 16:570.
55. Williams, J. 1971. The growth *in vitro* of killer particles of *Paramecium aurelia* and the axenic culture of this protozoan. *J. Gen. Microbiol.* 68:253.
56. van Wagtendonk, W. J., Clark, J. A. D., and Godoy, G. A. 1963. The biological status of lambda and related particles in *Paramecium aurelia. Proc. Natl. Acad. Sci. U.S.A.* 50:835.
57. Gibson, I. 1973. Transplantation of killer endosymbionts in *Paramecium. Nature* 241:127.
58. Koizumi, S. 1974. Microinjection and transfer of cytoplasm in *Paramecium. Exp. Cell Res.* 88:74.
59. Soldo, A. T. 1963. Axenic culture of *Paramecium*: Some observations on the growth behavior and nutritional requirements of a particle-bearing strain of *P. aurelia* 299 lambda. *Ann. N.Y. Acad. Sci.* 108:380.
60. Balsley, M. 1967. Dependence of the kappa particles of stock 7 of *Paramecium aurelia*, on a single gene. *Genetics* 56:125.
61. Preer, J. R. Jr., et al. 1971. Isolation and composition of bacteriophage-like particles from kappa of killer paramecia. *Mol. Gen. Genet.* 111:202.
62. Preer, J. R. Jr. and Preer, L. B. 1967. Virus-like bodies in killer paramecia. *Proc. Natl. Acad. Sci. U.S.A.* 58:1774.
63. Quackenbush, R. L. 1977. Phylogenetic relationships of bacterial endosymbionts of *Paramecium aurelia*: Polynucleotide sequence relationships among members of *Caedobacter. J. Bacteriol.* 129:895.
64. Quackenbush, R. L. 1978. Genetic relationships among bacterial endosymbionts of *Paramecium aurelia*: Deoxyribonucleotide sequence relationships among members of *Caaedobacter. J. Gen. Microbiol.* 108:181.
65. Preer, J. R. Jr. 1974. The intracellular bacteria of *Paramecium*. In *Bergey's manual of determinative bacteriology*, ed. R. E. Buchanan and N. E. Gibbons, 382. 8th ed. Baltimore: William & Wilkins.
66. Preer, J. R. Jr., Siegel, R. W., and Stark, P. S. 1953. The relationship between kappa and paramecin in *Paramecium aurelia. Proc. Natl. Acad. Sci. U.S.A.* 39:1228.
67. Jurand, A., Rudman, B. M., and Preer, J. R. Jr. 1971. Prelethal effects of killing action by stock 7 of *Paramecium aurelia. J. Exp. Zool.* 177:365.
68. Mueller, J. A. 1963. Separation of kappa particles with infective activity from those with killing activity, and identification of the infective particles in *Paramecium aurelia. Exp. Cell Res.* 30:492.
69. Tallan, I. 1959. Factors involved in infection by the kappa particles in *P. aurelia*, syngen 4. *Physiol. Zool.* 32:78.
70. Dilts, J. A. 1976. Covalently closed, circular DNA in kappa endosymbionts of *Paramecium. Genet. Res.* 27:161.
71. Quackenbush, R. L. 1983. Plasmids from bacterial endosymbionts of hump-killer paramecia. *Plasmid* 9:298.
72. Dilts, J. A. and Quackenbush, R. L. 1986. A mutation in the R body coding sequence destroys expression of the killer trait in *P. tetraurelia. Science* 232:641.

73. Quackenbush, R. L. and Burbach, J. A. 1983. Cloning and expression of DNA sequences associated with the killer trait of *Paramecium tetraurelia* stock 47. *Proc. Natl. Acad. Sci. U.S.A.* 80:250.
74. Kanabrocki, J. A., Quackenbush, R. L., and Pond, F. R. 1989. Organization and expression of genetic determinants for synthesis of type 51 R body-producing bacteria. *Microbiol. Rev.* 53:25.
75. Pond, F. R., et al. 1989. R-body-producing bacteria. *Microbiol. Rev.* 53:25.
76. Lalucat, J., et al. 1979. R bodies in newly isolated free-living hydrogen-oxidizing bacteria. *Arch. Microbiol.* 121:9.
77. Favinger, J., Stadtwald, R., and Gest, H. 1989. *Rhodospirillum centenum*, sp. nov., a thermotolerant cyst-forming anoxygenic photosynthetic bacterium. *Antonie van Leeuwenhoek* 55:291.
78. Siegel, R. W. 1954. Mate-killing in *P. aurelia*. *Physiol. Zool.* 27:89.
79. Beale, G. H. and Jurand, A. 1966. Three different types of mate-killer (mu) particles in *Paramecium aurelia* (syngen 1). *J. Cell Sci.* 1:31.
80. Levine, M. 1953. The diverse mate-killers of *P. aurelia*, variety 8: Their interrelations and genetic basis. *Genetics* 38:561.
81. Beale, G. H. and Jurand, A. 1960. Structure of the mate-killer (mu) particles in *Paramecium aurelia*, stock 540. *J. Gen. Microbiol.* 23:243.
82. Preer, L. B. et al. 1972. The classes of kappa in *Paramecium aurelia*. *J. Cell Sci.* 11:581.
83. Schneller, M. V. 1958. A new type of killing action in a stock of *Paramecium aurelia* from Panama. *Proc. Indiana Acad. Sci.* 67:302.
84. Schneller, M. V. 1962. Some notes on the rapid lysis type of killing found in *Paramecium aurelia*. *Am. Zool.* 2:446.
85. Sonneborn, T. M., Mueller, J. A., and Schneller, M. V. 1959. The classes of kappa-like particles in *Paramecium aurelia*. *Anat. Rec.* 134:642.
86. Jurand, A. and Preer, L. B. 1969. Ultrastructure of flagellated lambda symbionts in *Paramecium aurelia*. *J. Gen. Microbiol.* 54:359.
87. Dippell, R. 1950. Mutation of the killer cytoplasmic factor in *P. aurelia*. *Heredity* 4:165.
88. Preer, J. R. Jr. 1948. A study of some properties of the cytoplasmic factor, "kappa," in *Paramecium aurelia*, variety 2. *Genetics* 33:349.
89. Görtz, H.-D. 1987. Different endocytobionts simultaneously colonizing ciliate cells. *Ann. N.Y. Acad. Sci.* 503:261.
90. Fokin, S. I., et al. 1996. *Holospora* species infecting the nuclei of *Paramecium* appear to belong into two groups of bacteria. *Eur. J. Protistol.* 32(Suppl. 1):19.
91. Beale, G. H., Jurand, A., and Preer, J. R. Jr. 1969. The classes of endosymbiont of *Paramecium aurelia*. *J. Cell Sci.* 4:65.
92. Fokin, S. I. 2000. Host specificity of *Holospora* and its relationships with *Paramecium* phylogeny. *Jpn. J. Protoz.* 33:94.
93. Görtz, H.-D. and Wiemann, M. 1989. Route of infection of the bacteria *Holospora elegans* and *Holospora obtusa* into the nuclei of *Paramecium caudatum*. *Eur. J. Protistol.* 24:101.
94. Hafkine, W. M. 1890. Maladies infectueuses des paramécies. *Ann. Inst. Pasteur* 4:148.
95. Wichterman, R. 1985. *The biology of paramecium*, 1. 2nd ed. New York: Plenum Press.
96. Petschenko, B. 1911. *Drepanospira mülleri* n.g.n. sp. parasite des paraméciums, contribution à l'étude de la structure des bactéries. *Arch. Protistenkd.* 22:248.
97. Fokin, S. I. 1993. Bacterial endosymbionts of ciliates and their employment in experimental protozoology. *Cytologia* (Russ.) 35:59.
98. Fokin, S. I. and Ossipov, D. V. 1986. *Pseudocaedibacter glomeratus* sp. n.—The symbiont of the cytoplasm of *Paramecium pentaurelia*. *Cytologia* (Russ.) 28:1000.

99. Skoblo, I. I., Lebedeva, N. A., and Rodionova, G. V. 1996. The formation of the Paramecium-Holospora symbiotic system: A study of the compatibility of different symbiont isolates and host clones. *Eur. J. Protistol.* 32(Suppl. I):147.
100. Görtz, H.-D. 1996. Symbiosis in ciliates. In *Ciliates—Cells as organisms*, ed. K. Hausmann and P. C. Bradbury, 441. Stuttgart: Fischer.
101. Schmidt, H. J., et al. 1988. Characterization of *Caedibacter* endonucleobionts from the macronucleus of *Paramecium caudatum* and the identification of a mutant with blocked R-body synthesis. *Exp. Cell Res.* 174:49.
102. Kusch, J., et al. 2000. The toxic symbiont *Caedibacter caryophila* in the cytoplasm of *Paramecium novaurelia*. *Microb. Ecol.* 40:330.
103. Daugherty, R. M., et al. 2004. The nucleotide transporter of *Caedibacter caryophilus* exhibits an extended substrate spectrum compared to the analogous ATP/ADP translocase of *Rickettsia prowazekii*. *J. Bacteriol.* 186:3262.
104. Chao, P. K. 1953. Kappa concentration per cell in relation to the life cycle, genotype and mating type in *P. aurelia* variety 4. *Proc. Natl. Acad. Sci. U.S.A.* 39:103.
105. Gibson, I. and Beale, G. H. 1962. The mechanism whereby the genes M_1 and M_2 in *P. aurelia*, stock 540, control growth of the mate-killer (mu) particles. With an appendix by E. C. R. Reeve. Mathematical studies of mesosome distribution. *Genet. Res.* 3:24.
106. Reeve, E. C. R. 1962. Mate-killer (mu) particles in *Paramecium aurelia*: The metagon division hypothesis. *Genet. Res.* 3:328.
107. Yeung, K. K. 1965. Maintenance of kappa particles in cells recently deprived of gene *K* (stock 51, syngen 4) of *P. aurelia*. *Genet. Res.* 6:411.
108. Sonneborn, T. M. 1944. Exchange of cytoplasm at conjugation in *Paramecium aurelia*, variety 4. *Anat. Rec.* 89(Suppl.):49.
109. Gibson, I. and Sonneborn, T. M. 1964. Is the metagon an m-RNA in *Paramecium* and a virus in *Didinium*? *Proc. Natl. Acad. Sci. U.S.A.* 52:869.
110. Sonneborn, T. M. 1965. The metagon: RNA and cytoplasmic inheritance. *Am. Nat.* 49:279.
111. Beale, G. H. and McPhail, S. 1967. Some additional results on the maintenance of kappa particles in *Paramecium aurelia* (stock 51) after loss of the gene *K*. *Genet. Res.* 9:369.
112. Byrne, B. 1969. Kappa, mu, and the metagon hypothesis in *P. aurelia*. *Genet. Res.* 13:197.
113. Fisher, R. A. 1936. Has Mendel's work been rediscovered? *Ann. Sci.* 1:115.
114. Brenner, S., Jacob, F., and Meselson, M. 1961. An unstable intermediate carrying information from genes to ribosomes for protein synthesis. *Nature* 190:576.
115. Nass, M. M. K. and Nass, S. 1963. Intramitochondrial fibres with DNA characteristics. *J. Cell Biol.* 19:593.
116. Margulis, L. 1993. *Symbiosis in cell evolution*, 1. 2nd ed. New York: Freeman.
117. Beale, G. H. 1969. A note on the inheritance of erythromycin resistance in *P. aurelia*. *Genet. Res.* 14:341.
118. Knowles, J. K. C. 1974. An improved microinjection technique in *P. aurelia*: Transfer of mitochondria conferring erythromycin-resistance. *Exp. Cell Res.* 88:79.
119. Beale, G. H., Knowles, J. K. C., and Tait, A. 1972. Mitochondrial genetics in *Paramecium aurelia*. *Nature* (London) 235:393.
120. Maki, R. A. and Cummings, D. J. 1977. Characterization of mitochondrial DNA from *P. aurelia* with Eco R1 and Hae II restriction endonucleases. *Plasmid* 1:106.
121. Beale, G. H. 1973. Genetic studies on mitochondrially inherited mikamycin resistance in *P. aurelia*. *Mol. Gen. Genet.* 127:241.
122. Knowles, J. K. C. and Tait, A. 1972. A new method for studying the genetic control of specific mitochondrial proteins in *Paramecium aurelia*. *Mol. Gen. Genet.* 117:53.
123. Adoutte, A. and Beisson, J. 1972. Evolution of mixed populations of genetically different mitochondria in *Paramecium aurelia*. *Nature* (London) 235:393.
124. Adoutte, A. 1977. La génétique des mitochondries chez Paramécies. Thèse de Doctorat d'état, Paris XI Centre Orsay.

125. Tait, A. 1968. Genetic control of beta-hydroxbutyrate dehydrogenase in *Paramecium aurelia*. *Nature* (London) 219:941.
126. Tait, A. 1970. Genetics of NADP isocitrate dehydrogenase in *Paramecium aurelia*. *Nature* (London) 225:181.
127. Tait, A. 1970. Enzyme variation between syngens in *Paramecium aurelia*. *Biochem. Genet.* 4:461.
128. Ruiz, F. and Adoutte, A. 1978. Selection and characterization of nuclear mutations affecting mitochondria in *Paramecium tetraurelia*. *Mol. Gen. Genet.* 162:1.
129. Beale, G. H. and Knowles, J. K. C. 1976. Interspecies transfer of mitochondria in *P. aurelia*. *Mol. Gen. Genet.* 143:197.
130. Sainsard, A., Clarise, M., and Balmefrezal, M. 1974. A nuclear mutation affecting structure and function of mitochondria in *Paramecium*. *Mol. Gen. Genet.* 130:113.
131. Sainsard, A. 1975. Mitochondrial suppressor of a nuclear gene in *Paramecium*. *Nature* (London) 257:312.
132. Sainsard-Chanet, A. 1978. A new type of mitochondrial mutation in *Paramecium*. *Mol. Gen. Genet.* 159:117.
133. Sainsard-Chanet, A. 1979. Etude génétique et physiologique des interactions entre une mutation nucléaire et ses supresseurs mitochondriaux chez la paramecie. Thesis, Université Paris, Sud.
134. Sainsard-Chanet, A. and Knowles, J. 1979. Restoration of nucleo-mitochondrial compatibility in *Paramecium*. *Genetics* 93:833.
135. Tait, A. 1972. Altered mitochondrial ribosomes in an erythromycin resistant mutant of *Paramecium*. *FEBS Lett.* 24:117.
136. Cummings, D. L., Goddard, J. M., and Maki, R. A. 1976. Mitochondrial DNA from *P. aurelia*. In *The genetic function of mitochondrial DNA*, ed. C. Saccone and A. M. Kroon, p. 119. Amsterdam: North-Holland.
137. Cummings, D. L., Pritchard, A. E., and Maki, R. A. 1979. Restriction enzyme analysis of mitochondrial DNA from closely related species of *Paramecium*. In *Proceedings of the 1979 Symposium on Molecular and Cellular Biology*. Vol. 15, *Extrachromosomal DNA*, ed. D. L. Cummings et al., 35. New York: Academic Press.
138. Tait, A., et al. 1976. The study of the genetic function *Paramecium* mitochondrial DNA using species hybrids. In *Genetics and biogenesis of chloroplasts and mitochondria*, ed. T. Bucher et al., 569. Amsterdam: Elsevier.
139. Tait, A., Knowles, J. K. C., and Hardy, J. C. 1976. The genetic control of mitochondrial proteins. In *The genetic function of mitochondrial DNA*, ed. C. Saccone and A. M. Kroon, 131. Amsterdam: Elsevier.
140. Pritchard, A. E., et al. 1990. Nucleotide sequence of the mitochondrial genome of *Paramecium*. *Nucl. Acids Res.* 18:173.
141. Gray, M. W., et al. 1998. Genome structure and gene content in protist mitochondrial DNAs. *Nucl. Acids Res.* 26:865.
142. Gray, M. W., Lang, G. F., and Burger, G. 2004. Mitochondria of Protista. *Annu. Rev. Genet.* 38:477.
143. Seraphin, B., et al. 1987. Construction of a yeast strain devoid of mitochondrial introns and its use to screen nuclear genes involved in mitochondrial splicing. *Proc. Natl. Acad. Sci. U.S.A.* 84:6810.
144. Cummings, D. L. 1992. Mitochondrial genomes of the ciliates. *Int. Rev. Cytol.* 141:1.
145. Fox, T. D. 1987. Natural variation in the genetic code. *Annu. Rev. Genet.* 21:67.
146. Osawa, S., et al. 1992. Recent evidence for evolution of the genetic code. *Microb. Rev.* 56:229.
147. Tourancheau, A. B., et al. 1995. Genetic code variations in the ciliates: Evidence for multiple and single events. *EMBO J.* 14:3262.

7 Determination of i-Antigens

7.1 INTRODUCTION

Injection of paramecia into rabbits produces antiserum that reacts with immobilization antigens, or i-antigens, of *Paramecium*. In this chapter we will see how the story of the antigens has developed over the years. Now that the proteins and the genes that code for them have been identified and sequenced, we have a different perspective on the i-antigens, and a different way of approaching the problems today compared with what was done formerly. But when the story began we were primarily dependent on the techniques of genetics and immunology for our information. For instance, the differences between the B protein in stock 51 and that in stock 29 were impossible to determine at that time because no existing serum was able to tell one from the other. But now we can simply look at the gene sequences and compare them. In the following pages we try to present the development of our knowledge of the i-antigens in the way it actually occurred, that is, chronologically, for research on the antigens began long before molecular biology became important and James Watson had even been born.

The immobilization antigens of *Paramecium aurelia* are exceptionally valuable materials for the study of certain basic genetic and epigenetic problems. These substances are sometimes denoted simply as surface proteins, but will be referred to here as i-antigens, since there are few other surface proteins, apart from the i-antigens, in *Paramecium*. We will concentrate attention on the i-antigens here and ignore other surface proteins. The existence of large numbers of different variants of the surface antigens in different wild stocks of *Paramecium* provides us with a virtually unlimited amount of variation for genetic analysis.

The genetic system that this material has revealed is one in which a given genic constitution can undergo a process bringing about a switch from a state in which one gene is expressed, to another state favoring expression of another gene. These changes have proved to be cytoplasmically inherited and cannot be explained by simple Mendelian means. Instead, they are a mixture of genetic and epigenetic changes. As such, they represent a case of cellular differentiation similar to that seen in embryonic development in higher organisms, and provide us with an opportunity to investigate a case of cellular differentiation in *Paramecium*.

In addition to providing valuable materials for the study of certain basic genetic and epigenetic problems, the *Paramecium* i-antigen system also provides us with a model for understanding some aspects of antigenic variation in parasitic organisms, such as malaria parasites and trypanosomes, a matter of great medical importance, though there are important differences between the *Paramecium* system and that present in these other protists.

We will first outline the main facts of the *Paramecium* antigen system, before going on to discuss the details of various genetic and epigenetic, chemical, and molecular biological aspects. This review will no doubt seem unusual to some present-day readers because it devotes considerable attention to researches involving, for the most part, techniques of classical (i.e., Mendelian) genetics, on which our present knowledge of the system is founded. This may need to be clarified, especially as this Mendelian analysis involves certain considerations, for example, formation of the macronucleus from the micronucleus, that are not straightforward. In addition, some more modern research on the subject, using molecular techniques and certain epigenetic phenomena, will be discussed.

The account will be confined mainly to the *P. aurelia* complex of *Paramecium*, though other morphospecies of *Paramecium*, and other ciliates and bacteria that exhibit related phenomena, have also been studied. The other species of *Paramecium*, apart from the *P. aurelia* group, are *P. caudatum*[1] and *P. multimicronucleatum*,[2] as well as other ciliate genera, such as *Tetrahymena* (see Doerder,[3] Sonneborn,[4] Williams et al.,[5] Doerder and Berkowitz,[6] and Love et al.[7]).

7.2 THE IMMOBILIZATION TEST AND SEROTYPES

It has been known for many years that a few injections of suspensions of *Paramecium* into the blood of a rabbit result in the development of antibodies having characteristic effects on living organisms of the same immobilization type as those injected.[8] Various methods of carrying out immunization with *Paramecium* have been described. Clear results can be obtained by the following procedure: A few thousand paramecia, that is, those contained in about 10 ml of a culture of *P. tetraurelia* in grass or lettuce media, are concentrated by centrifugation, and a homogenate is prepared by forcing the cells through a fine injection needle, or by freezing and thawing them. Four to six intravenous injections of the homogenate are made into the ear of a rabbit, at twice-weekly intervals, and after another week or two samples of blood are withdrawn. The blood withdrawn is allowed to clot, and the serum collected and diluted into Ringer's solution. Samples of the serum can be tested immediately for their immobilizing effect on paramecia, or can be stored in a refrigerator or deep freeze until required.

To test for the effect of the antisera on living paramecia, a series of dilutions of the antiserum is made in dilute Ringer's solution and added to culture fluid containing a few paramecia, or just a single cell, in a glass depression slide, which is then examined with a low-power binocular microscope. It is commonly found that antiserum prepared as above and diluted about 1:800 is capable of immobilizing in 2 hours paramecia belonging to the same antigenic type as that used for the immunization of the rabbit.

As pointed out by Finger,[9] treatment of paramecia with immobilization antisera does not produce agglutination of the organisms like that caused by treatment of bacteria with immune antisera against bacteria. Furthermore, complement is not required for immobilization of *Paramecium* by immune antisera, and it is not necessary to absorb the antisera with nonhomologous antigens before doing the immobilization tests, because of the high specificity of the immobilization reaction and the fact that the i-antigens are such good antibody inducers. When paramecia are placed in a solution containing antiserum bearing specific immobilizing antibodies, the cilia clump together and the normal swimming motion of the organism is inhibited. Electron microscope studies have shown that the clumping of the cilia is caused by fusion of the plasma membranes at their tips[10] and involves cross-linking between immunoglobulin (IgG) in the antiserum and i-antigen in the cell membrane. Monovalent antibody fragments prepared from a serum do not cause immobilization unless rabbit immunoglobulin (IgG) is added.[10,11] Moreover, monoclonal antibodies against *Paramecium*, although they are absorbed on to the surface of the cells, do not cause immobilization.[12]

By treatment of living paramecia with immobilizing antiserum conjugated with fluorescein or other fluorescent dyes, and observation of the treated cells by ultraviolet microscopy, visual evidence has been obtained that specific antibodies accumulate in large globules at the tips of the cilia. The stems of the cilia of living paramecia treated in this way appear to be free of antibody. However, if the paramecia are first fixed with osmic acid or gluteraldehyde before fluorescent antibody treatment, and the cilia as a consequence of fixation are not beating, no antibody molecules are swept up the sides of the cilia to accumulate as globules on the tips of the cilia under these conditions. Instead, a uniform fluorescence is seen in the pellicle and along the whole lengths of the cilia,[13] and it is clear that i-antigens are present over the whole external surface of *Paramecium*, including the pellicle and the cilia. Nonhomologous cells, that is, those cells of a different antigenic type from those used for preparation of the antiserum, do not absorb fluorescent antibody on any part of their surface and are not immobilized. However, they take up large amounts of fluorescent material in the food vacuoles if the paramecia are left in the solution for a sufficient time. There is no retention of nonhomologous antibodies in the cytoplasm of living cells.

Immobilization by specific antisera does not actually kill paramecia, for the cytoplasm can be seen to be circulating in the interior of cells even when immobilization is complete, and cells may recover motility after being immobilized, provided the treatment has not been very severe. The immobilization reaction is one involving substances on the external surface of *Paramecium*.[13–16] The possibility that the i-antigen is normally shed from the surface of *Paramecium* into the medium, though undoubtedly found, is difficult to prove, for if it is found in the medium in mass cultures, it might have originated from occasionally dying cells. On the other hand, if it is assayed in the medium from very small cultures where all cells can be accounted for, then it must be present in only small quantities. In any case, it is concluded that the immobilization antigen is a substance situated on the external surface of the paramecia, and, as will be described, it is a protein covering the cilia and pellicle, or constituting part of the cilia and pellicle. Even though it is not the only surface protein, it is present in relatively large quantities: the immobilization antigen is the most abundant surface protein in *Paramecium*, representing 25–30% of the ciliary protein,[17] or about 3.5% of the total protein of *P. tetraurelia*.[18] The thickness of the layer of the immobilization antigens covering the cells is about 17–25 nm.[16,19]

Since all living paramecia possess immobilization antigens, they are considered to be essential components of the cells. It has been suggested by Harumoto and Miyake[20] and Harumoto[21] that they are important in protecting paramecia from natural predators, such as *Didinium nasutum*, with different serotypes (see below) providing different degrees of protection against *D. nasutum*. Substances somewhat similar to the immobilization antigens of the *P. aurelia* complex are known in certain parasitic protozoa such as *Trypanosoma*, and in that case are thought to play a role in the defensive mechanisms developed by the host organism against the parasite. A second possible role for the i-antigens is that their abundance suggests that they merely act as a buffer against pH and metal ions in the external medium.[22] A third possibility is that they are involved in chemoresponses. This is suggested because the i-antigens, like many receptors and ion channel molecules, are anchored to the cell surface through linkage to glycosyl phosphatidylinositol (GPI).[23]

Several techniques have been used for measuring the concentration of the i-antigens of *Paramecium*. The amounts can be measured by determining the dilution of an antigen preparation needed to adsorb a standard quantity of antibody from a serum.[24] The antigens can also be measured by the technique of precipitation in agar gels.[25] Much of the genetic work has been done using precipitation in gels, which, as well as being technically very simple and quick, is both quantitative and very sensitive in regard to detection of fine differences between different cells. A third method, which may only be used in vivo, is simple but only approximate. This method is to measure the time necessary for complete immobilization of paramecia by a given concentration of specific antibody.[26]

A remarkable feature of the i-antigen system of *Paramecium* is that cells of any homozygous stock of the organism can exist in alternative states, called serotypes, which are denoted, for example, 51A, 51B, 51C, and so on, for variants of stock 51,[27] corresponding to the presence of different i-antigens, A, B, C, and so on, in paramecia. As will be described, genetic determination of the serotypes is controlled by series of genes, which are denoted by italicized capital letters with superscripts to indicate alleles, for example, A^{51}, B^{51}, C^{51}, and so forth. This notation is adopted here for reasons of consistency with standard genetic conventions. An alternative nomenclatural system for the i-antigens has been proposed by Allen et al.,[28] but has not gained acceptance.

Paramecia are extraordinarily variable in regard to their i-antigens. A given homozygous stock of *P. tetraurelia* is capable of giving rise to a dozen or more different serotypes, each of which is recognized by immobilization of the paramecia by a specific antiserum. Moreover, different stocks having different genetic constitutions can produce different sets of serotypes, as shown below. The genetic basis of this system was first investigated in the species *P. tetraurelia*,[27] and later in the species *P. primaurelia*.[29] While the almost universal rule is that only one antigen is present on the surface at a time, an exception was found in the case of 172D and 172M,[30] where cells reacted with both D and M antisera.

A few details will illustrate the basic genetic features of the i-antigen system. In their early work, Sonneborn and LeSuer[27] showed that a sample of a single homozygous stock (stock 51 of *P. tetraurelia*) had the capacity to form just four serotypes at that time, each reacting with a corresponding homologous antiserum (anti-A, anti-B, anti-C, and anti-D, respectively). Thus, starting with a pure clone of stock 51, *P. tetraurelia*, grown at 27°C with sufficient food for a growth rate of one fission per day,

TABLE 7.1
Titration of Four Antisera with Four Serotypes of Stock 51, *P. tetraurelia*

Paramecia	Antisera			
	anti-51A	anti-51B	anti-51C	anti-51D
Serotype 51A	1/800	1/50	1/100	1/1000
Serotype 51B	1/25	1/3200	1/200	1/200
Serotype 51C	1/3	1/6	1/1600	--
Serotype 51D	1/12.5	--	1/3	1/3200

The titres given indicate the dilution of antiserum necessary to immobilize paramecia of a given serotype in the standard time of 2 hrs.
From Sonneborn, T.M. and LeSuer, A., Antigenic characters in *Paramecium aurelia* (variety 4): Determination, Inheritance and Induced Mutation, *Amer. Nat.*, vol. 82, p. 75, Table IV, 1948, with permission of the University of Chicago Press.

the serotype 51A was produced. However, the same culture of a single pure stock could sometimes produce a few cells that were unaffected by diluted anti-A serum. If these exceptional cells were allowed to multiply in medium containing anti-51A antibodies, any accompanying A cells were eliminated, and a second serotype was produced by the surviving cells. This second type could be grown up and used for the preparation of a second antiserum that immobilized the second type of cells, but not the original A type. The second type was denoted 51B. By further work, two more types, C and D, were readily produced. The interactions between these four serotypes and the corresponding antisera are shown in Table 7.1.

Moreover, later work showed that stock 51 had the potentiality to form not only the four serotypes A, B, C, and D, but as many as eleven different serotypes, which were denoted A, B, C, D, E, G, H, I, J, N, and Q. Stock 29 could form at least eight types: A, B, C, D, F, G, H, and J. Stock 172 could form at least twelve types: A, B, C, D, E, F, G, J, M, N, O, and P.[31] Corresponding types in different stocks of *P. tetraurelia* were denoted with the same letter, and most were indistinguishable from each other. Sometimes corresponding types showed varying (but usually close) serological relationships, and were also designated by the same letter.[32,33] We now know from DNA sequencing studies (see below) that virtually all the serotypes are slightly different in their base sequences in different stocks. Therefore, it is customary to designate the serotype by adding the name of the stock, such as 51A, 29A, 51B, and 29B, irrespective of whether there are slight serological differences between the two (51A and 29A) or no difference (51B and 29B).

7.3 THE GENETIC SYSTEM OF THE I-ANTIGENS

7.3.1 EARLY WORK BY SONNEBORN: GENES, CYTOPLASM, AND ENVIRONMENT

Although the different serotypes of a given stock of *P. tetraurelia* may be formed under different conditions of temperature, growth rate, and so on, more than a single serotype can be maintained under exactly the same environmental conditions. Thus, each of the three serotypes 51A, 51B, and 51D can be formed and maintained indefinitely at 27°C in test-tube cultures of stock 51 containing sufficient food to produce one fission per day.[27,32] Contrary to the expectations of classical geneticists, these three intrastock variants were not found to be genically inherited, but were controlled by the cytoplasm. Thus, when cells of serotype 51A were allowed to conjugate with those of type 51B (Figure 7.1), the exconjugants deriving cytoplasm from the 51A parent produced clones of 51A cells, while exconjugants deriving cytoplasm from the 51B parent produced clones of 51B cells. The different serotypes were controlled by the cytoplasm rather than by different parental macronuclei, as shown by the finding that if a substantial amount of cytoplasm passed from one mate to another during conjugation, an abnormal state of affairs, exconjugants often changed. In the example shown in Figure 7.1, one exconjugant produced a mixed line that eventually segregated into pure A and B lines. These facts eliminate the hypothesis that control is effected by the parental macronuclei or genes and prove that the cytoplasm is responsible.

Other experiments, however, showed that nuclear (Mendelian) genes were, after all, also involved in the determination of the i-antigens. Thus, stocks 51 and 29, as indicated above, differed in that stock 29 was able to form the serotype 29F, while stock 51 was unable to form any corresponding F type. Crosses between stocks 29 and 51 and passage of the resulting F1s through autogamy gave rise to F2 families in which there was a 1:1 segregation of the ability or inability to form any type F, irrespective of the cytoplasmic parentage. Therefore, the ability or inability to form the antigen F was controlled by a pair of Mendelian factors, which will here be denoted F^{29} and F^o.

A second indication of genic variation between stocks 51 and 29 of *P. tetraurelia* concerns a small but distinct serological difference between the serotypes 51A and 29A, as seen in the titers of these two sera with homologous and heterologous sera. Sonneborn[32] studied conjugation between the two types 51A and 29A and allowed the F1 hybrids subsequently to pass through autogamy. He then found that in the ex-autogamous F2 thereby obtained, there was a 1:1 segregation of the two serotypes called 51A and 29A, indicating that the "specificity" of the A-type antigens was under the control of a pair of allelic genes designated A^{51} and A^{29}. Further, it was shown[34] that the specificity of another pair of serotypes, 51H and 29H, was likewise controlled by a pair of alleles, H^{51} and H^{29}, and these were at a different gene locus from that of the *A* alleles. Thus, in this early work, Sonneborn[32] showed that both genic and cytoplasmic factors were involved in the heredity of the i-antigens of *P. tetraurelia*.

The specificity alleles, H^{51} and H^{29}, were shown to be at the same gene locus as the null allele (H^o), which could not produce any H antigen. Allelism of specificity and null alleles were also shown for genes at the *A* and *F* loci by Reisner,[35] who also showed by Mendelian analysis that specificity and null alleles were at the same locus. Thus, it appears that each serotype is governed by a single gene that encodes the primary structure of the i-antigen characteristic of that serotype. This hypothesis was subsequently validated by biochemical analysis, showing that there is a specific gene for each serotype in a stock in *P. tetraurelia*[36,37] and in *P. primaurelia*.[38]

7.3.2 Environmental Control of the Cytoplasm

The genetics of i-antigen variation was studied not only in the species *P. tetraurelia*, but also in *P. primaurelia*.[29] The species *P. primaurelia* had previously been

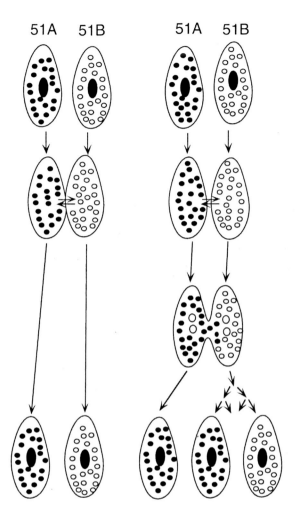

FIGURE 7.1 Conjugation between serotypes A and B, without and with cytoplasmic exchange. The figure shows normal conjugation on the left and cytoplasmic exchange on the right, depicted at several time points. The top row of four cells are just before conjugation. The black circles in the 51A cells represent the A cytoplasm, while the open circles represent the B cytoplasm. The second row of cells represents the time in conjugation when the micronuclei migrate across and fertilization occurs. Old macronuclear fragments are not shown. The third row of cells represents the two-anlagen stage, with cytoplasmic exchange occurring in the single pair represented on the right. If cytoplasmic exchange occurs in the opposite direction, it is usually ineffective in producing change. The bottom row of cells represents the final result of no cytoplasmic exchange (the left two cells) and after exchange (the right three cells). The right-most two cells undergo several fissions before they stabilize as pure types. Reprinted with permission of Cambridge University Press; modified from Beale, G. H., *The Genetics of* Paramecium aurelia (London: Cambridge University Press, 1954), 88, Figure 4.

considered by Sonneborn to be genetically more conventional than the species *P. tetraurelia*, for he had thought at one time that *P. primaurelia*, unlike *P. tetraurelia*, behaved according to Mendelian rules, while *P. tetraurelia* was thought to exhibit cytoplasmic heredity. Sonneborn's views on genetics at that time were based on his early work on the kappa particles in *P. tetraurelia*, and on the cytoplasmic inheritance of the difference between serotypes in the species *P. tetraurelia*, which is a member of the mating type group B. In *P. primaurelia*, which Sonneborn classified as a group A species, Mendelian heredity was the rule.[39,40]

However, after further investigation of i-antigen variation in *P. primaurelia*, a fundamentally similar genetic situation to that of *P. tetraurelia*, involving both genic and cytoplasmic determinants in *P. primaurelia*, as described below, was shown to exist.[29] The following is a summary of the genetics of the antigen system in *P. primaurelia*, according to Beale.[29,41] In species 1, many stocks were found to be capable of giving rise to three serotypes, which were denoted S, G, and D, and were formed in cells grown at low, medium, and high temperatures (18, 25, and 29°C or higher), respectively (though the S types were usually produced not only by low temperatures but also by slow growth rates). By cross-testing sera and stocks of *P. tetraurelia* and *P. primaurelia*, the medium-temperature type was found to be similar to serotype G in *P. tetraurelia* and was named type G. The high-temperature type was similar to type D, so it became D. The low-temperature type proved to have no cross-reactions with any of the types in *P. tetraurelia*, and it was designated serotype S.

Genetic analysis of antigenic variations in the species *P. primaurelia* seemed at first to indicate that there was no cytoplasmic inheritance but only simple Mendelism. Indeed, as shown earlier, on the basis of crosses between stocks 60 and 90 followed by autogamy, serotypes D and B are governed by two independent Mendelian genes.[29,42] After autogamy, each of the two homozygous parental and each of the two homozygous recombinant clones were produced, for a total of four types all equal in number. Since the four types are equal in number, no linkage is shown. See Table 3.1. Therefore, stock 60 has the genotype d^{60}/d^{60} g^{60}/g^{60} and stock 90 has the genotype d^{90}/d^{90} g^{90}/g^{90}. The genes in the species *P. primaurelia*, unlike those of other species, are expressed as small letters. The phenotypes of the F1s and F2s in *P. primaurelia* were found to be expressed after five fissions following the change of genotype produced by conjugation or autogamy. This five-fission delay was a consequence of the time required for development and expression of new macronuclear genes following nuclear reorganization at autogamy or conjugation. It is called cytoplasmic lag.

Unlike the cross of 90 × 60, the cross of 60 × 61 differed at all three loci, d, g, and s. The F2 clones derived from a cross between stocks 60 and 61, and grown at low, medium, and high temperatures, yielded the expected eight different combinations of two alleles at each of three loci (s, g, and d) in equal numbers, showing independence of the three alleles.

The apparent role of the cytoplasm in the i-antigen system in *P. primaurelia*, which had been so striking in *P. tetraurelia* (as shown above), was demonstrated in *P. primaurelia* by crosses of serotypes 90G (stable at 25°C) to 60D (stable at 29°C and higher). See Figure 7.2. Such crosses can be made at 27°C because there is a lag of up to fifty fissions after the cells are changed in temperature before the serotypes actually switch. When the cross 90G × 60D was made at 27°C, it was found that after

Determination of i-Antigens

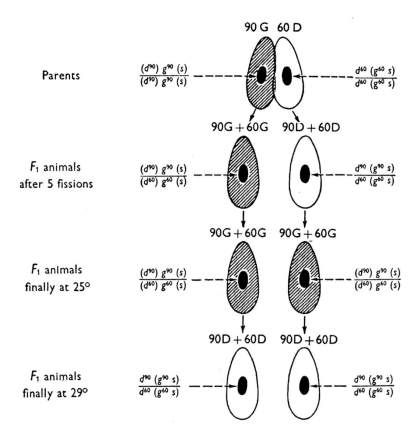

FIGURE 7.2 Results of a cross between 90G and 60D. See text for explanation. Reprinted with permission of Cambridge University Press; modified from Beale, G. H., *The Genetics of Paramecium aurelia* (London: Cambridge University Press, 1954), 95, Figure 6.

five fissions, those F1 paramecia that were the cytoplasmic progeny of the 90G parent developed a mixture of the antigens 90G and 60G, and those receiving cytoplasm from the 60D parent developed a mixture of the antigens 90D and 60D. If growth of the F1 cells was continued at 29°C, all the progeny eventually came to have the antigens 90D + 60D, no matter from which of the parent cells the cytoplasm had been derived, while continued growth at 25°C eventually produced only 90G + 60G types. These results were interpreted to mean that the cytoplasm of the 90G cells was of such a nature as to permit the expression not only of the gene g^{90} but also of g^{60} (and probably also of all other g alleles), while the cytoplasm of the phenotype 60D cells favored the expression of the two alleles d^{90} and d^{60}.[29] This property of the cytoplasm was transmitted from cell to cell for varying lengths of time, but could be changed eventually by varying the temperature.

Thus, the finding of Sonneborn and LeSuer[27] that antigenic variations within a stock of *P. tetraurelia* were apparently cytoplasmically controlled, but that particular antigens could only be developed in the presence of certain genes, was confirmed in

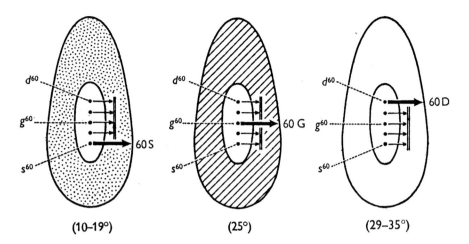

FIGURE 7.3 Interaction between cytoplasm and genes in stock 60 of *P. primaurelia*. See text for explanation. Reprinted with permission of Cambridge University Press; modified from Beale, G. H., *The Genetics of* Paramecium aurelia (London: Cambridge University Press, 1954), 96, Figure 7.

P. primaurelia.[29] This can be expressed more explicitly by saying that the specificity of the i-antigens is controlled by genes at different loci, with a number of alleles at each locus, and the expression of the genes at each locus is governed by what is called the cytoplasmic state.[29] See Figure 7.3. This figure shows that the cytoplasm becomes different at different temperatures. The cytoplasm at 10–19° blocks all the genes but s^{60} (see large bold arrow in the figure at 10–19°). Similarly, all the genes but g are blocked at 25° and all but d are blocked at 29–35°.

In his early papers, Sonneborn adopted a hypothesis based on hypothetical genetic units called plasmagenes as determinants of the i-antigens.[27] Moreover, subsequent genetic analysis showed that there were apparently no genic differences between types 51A and 51B, and this at first seemed to support the idea of plasmagenes. However, he later abandoned this interpetation. At this point it is interesting to recall a suggestion of Delbrück,[43] noted in Chapter 1, which he put forward as an alternative to the plasmagene hypothesis, formerly favored by Sonneborn. According to Delbrück's hypothesis, *Paramecium* cells might be considered a system showing flux equilibria and capable of passing through a series of steady states. It was speculated by Delbrück at that time that the antigens of *Paramecium* might be synthesized as a result of the activity of a different series of competing chains of enzyme-controlled reactions, and that the intermediate products of one reaction might act as inhibitors of competing reactions. However, this hypothesis is now seen to be inapplicable in the form proposed by Delbrück, because of later knowledge of the molecular biology of protein synthesis, not understood at the time when the genetic analysis of i-antigen variation in *Paramecium* was first made. Moreover, it was also unknown at that time that the i-antigens of *Paramecium* were proteins, and their formation was due to the synthesis of these proteins. Nevertheless, this hypothesis of Delbrück's is still considered a possibility, because it might be applied to some other system that in turn controls serotype stability. It occupies an important place in the study of the i-antigens, for it showed how mutual exclusion might be explained on the basis of a series of competing chemical reactions, instead of self-reproducing particles.

Later hypotheses, instead of operating on the basis of inhibition, assume that all pathways are turned off until one is activated. See, for example, the paper by Monod and Jacob,[44] who developed a whole series of models based on regulator genes and operons. It is unlikely, however, that any of these schemes will turn out to be the correct one operating in mutual exclusion of the i-antigen genes in view of recent experiments that show that the protein itself is involved in the exclusion phenomenon (see below).

As stated above, the type of hypothetical cytoplasmic state formed by a given stock of *P. tetraurelia* or *P. primaurelia* at a given temperature, and favoring expression of particular antigen-determining genes, is in its turn controlled by the environment in which the organisms are grown. Among the environmental factors involved are temperature, growth rate, quantity and kind of food, salinity, and pH. All these factors may cause switches from one serotype to another, in stocks having a particular genotype. The conditions favoring the expression of allelic genes at a particular antigen-determining gene locus cannot, however, be exactly specified, because of the many unknown factors involved. In spite of these complexities, it has been stated (see Preer[31]) that in stock 51 of *P. tetraurelia*, serotype H is most stable at 12°C, serotype B is most stable at 19°C, and serotype A is most stable at temperatures above 31°C. Detailed data on the environmental conditions affecting the induction, stabilization, and switching of all the serotypes known in stock 51 (*P. tetraurelia*) have been summarized by Austin[45-47] and Austin, Widmayer, and Walker.[48]

One phenomenon much studied in the early work on the i-antigen system of *P. tetraurelia* was the effect of sublethal doses of immobilizing antibodies on change of serotype of the treated cells. It had long been known[49] that a culture of paramecia, after treatment with dilute immobilizing antiserum, sometimes gives rise to cells that are resistant to the action of that serum. Some early workers, including perhaps even Sonneborn himself at one time (see Sonneborn[50]), thought that these phenomena gave some support to Lamarckian ideas. Careful observations have made it clear, however, that the induced switching of the antigenic changes of *Paramecium* consequent upon treatment with homologous antiserum was due to changes in gene expression, rather than changes in gene structure induced by some kind of adaptive or Lamarckian response to an unfavorable environment.

Serum-induced switches, for example, 51D treated with anti-51D and forming 51B, in *P. tetraurelia*, were analyzed by Sonneborn and LeSuer[27] and Beale.[51] It was found by the last-named author that the paramecia, in work of this kind, on first recovering from partial immobilization by weak anti-51D serum, contained the same antigen (i.e., type 51D) as that present before treatment, but in lesser amount, and that only after a subsequent period of about 15 hours' growth did a different (i.e., type 51B) antigen appear and eventually replace the old one (51D). The immobilization treatment therefore did not directly change the 51D antigen by some mutagen-like process, but changed the behavior of the cell in such a way that the latter, on further growth, expressed a new type of antigen, 51B.

However, paramecia that have been immobilized by a specific antiserum and subsequently regained normal motility, do not always switch to another serotype. Whether they do so or not depends on many factors, such as the type of antigen present in the starting material, the environmental conditions before and after application of the antiserum, and the genotype of the treated cells. Some serotypes change readily, others do so less readily, while some become even more stable as a result of homologous serum treatment than they would have been had there been no serum treatment at all.[52]

It should be added here that experienced workers have found that these phenomena sometimes show baffling irregularities, for example, an expected serotype switch may not appear at all or may appear only after a very long interval of time.[9] To summarize, the cytoplasm possesses a hereditary property, that is, there is a tendency for a given serotype to favor further production of itself, as shown by the perpetuation of the different serotypes in species 4 (*P. tetraurelia*) serotypes.

7.3.3 ALLELIC VARIANTS OF ANTIGEN-DETERMINING GENES

It has been shown above with both *P. tetraurelia* and *P. primaurelia* that there are alleles at several different genetic loci that determine the specificity of the i-antigens. We now need to consider how many different gene loci (s, g, d, etc.) there are in *P. primaurelia*, how many alleles at each locus (g^{60}, g^{90}, etc.).

As already stated, stock 51 of *P. tetraurelia* is capable of forming at least eleven or twelve different serotypes, some (or possibly all) of which are controlled by different alleles. It is therefore reasonable to assume that there are at least eleven or twelve gene loci in *P. tetraurelia* controlling these serotypes, though genetic analysis of variants of all these types by Mendelian experiments has not been done. According to Sonneborn,[53] at the A locus three distinguishable alleles, here denoted A^{29}, A^{32}, and A^{51}, have been shown to exist, and at the H locus, there are four alleles, H^{29}, H^{32}, H^{51}, and H^{169}. Similar evidence is available for most of the serotypes that stocks of *P. tetraurelia* are capable of forming, and which have been analyzed by Mendelian methods. Therefore, in *P. tetraurelia* there is proof of the existence of at least six i-antigen-determining gene loci with at least three or four alleles at each locus. However, detailed study of the numbers of alleles at individual loci is difficult in *P. tetraurelia* because of the relatively slight immunological differences produced by different alleles at a given locus in this species.

The situation in this regard is easier to analyze in *P. primaurelia*, where the different allelic variants of the antigens are more divergent than in *P. tetraurelia*, and indeed in some cases different allelic types in *P. primaurelia* appear to be totally unrelated when compared by the immobilization test. An attempt was made (see Beale,[42] p. 105) to classify forty-two stocks of *P. primaurelia* in regard to their allelic antigens of the different types formed by different stocks to the available set of each of the s, g, and d loci. This classification was based on the use of differentiating antisera. The results are summarized in Table 7.2, which indicates that at least four different s alleles, nine g alleles, and eight d alleles are clearly distinguishable in the available material. It was also shown that, as expected, crosses between any two homozygous stocks gave only the two allelic parental variants in their progeny.

The classification shown in Table 7.2 is based on the immunological reactions of the different stocks of *P. primaurelia* to the antisera available at the time when the study was made. If more antisera with finer powers of discrimination had been

TABLE 7.2
Wild Stocks of *P. primaurelia* and the Antigen Types at the S, G, and D loci

Stock No.	Geographical Origin	Antigen types		
		S	G	D
B,N,O,P,T	Woodstock, Md.	60	90	90
J	J. H. University Baltimore	61	90	90
R	Baltimore, Maryland	61	90	90
S	Cold Spring Harbor, N.Y.	60	61	41
Z	Stanford, California	60	61	60
33	Baltimore, Maryland	143	33	33
41	Atlanta, Georgia	60	41	41
43	Stanford, California	60	61	60
60	Burlingon, Virginia	60	60	60
61	Woods Hole, Mass.	61	61	90
62	Woods Hole, Mass.	60	60	90
90	Bethayers, Pennsylvania	60	90	90
103	Philadelphia, Penn.	60	90	103
119	Gwynedd, Pennsylvania	60	90	90
129	Fort Lauderdale, Florida	61	60	41
143	Falkirk, Scotland	143	41	143
144	Chantilly, France	60	41	90
145	Paris, France	60	60	145
147	Sendai, Japan	60	41	60
153	New Haven, Connecticut	60	90	90
156	New Haven, Connecticut	60	156	60
168	Sendai, Japan	61	168	90
171	Yamagata, Japan	143	41	60
175	Lake Titicaca, Peru	60	41	175
177	Santiago, Chile	61	61	60
180	Tokyo	143	61	60
182	Sano-si, Japan	60	41	90
183	Tokyo	60	41	90
513	Chantilly (1953)	60	60	90
514	Chantilly (1953)	60	60	145
515	Chantilly (1953)	60	60	145
516	Chantilly (1953)	60	60	90
520	Paris, France	60	41	41
523	Luzern, Switzerland	60	61	60

The table gives the stock and its location in Columns 1 and 2. Paramecia were cultured at different temperatures, allowing expression of S, G, and D serotypes. The numbers in column 1 refer to the stock used to make the serum. These classifications depend on the specificity of the sera used to make the tests. We also know now that the DNA sequences of all these types may be slightly different.

From Beale, G.H., *The Genetics of* Paramecium aurelia, p. 105, Table 19, 1954. Reprinted with permission of Cambridge University Press.

available, further subdivision of the antigenic categories might have been possible. Indeed, after study of the chemical differences between the allelic and nonallelic variants, it became apparent that an almost limitless amount of variation, sometimes too delicate to be detected by the immobilization test alone, is present in nature. For example, by two-dimensional chromatography and electrophoresis of tryptic digests of antigen preparations, it was found by Jones and Beale[38] that five stocks of *P. primaurelia*, namely, stocks 60, 178, 181, 521, and 535, which had previously been classified by the immobilization test as members a single group denoted 60D (see Table 3.1), could all be distinguished by variations in the two-dimensional fingerprints of certain peptides, though detailed proof by formal Mendelian analysis for the existence of all five allelic variants was not carried out. Data on base sequences suggest that virtually all i-antigens in different stocks may be different.

There is usually no obvious connection between a particular allelic variant as here designated and the geographical sites at which that variant was located. At a single geographic site, several stocks may be isolated bearing quite different alleles affecting a given serotype, while at widely separated geographic sites, apparently similar alleles may occur. As can be seen from Table 7.2, several different alleles occur at each locus, but collections of paramecia from different continents may yield assortments of apparently similar serotypes. Thus, serotype 60G occurs in the United States and in France in *P. primaurelia*. In another species of *Paramecium*, *P. novaurelia*, there are three different alleles at the G locus and six different alleles at a second locus denoted X, all found in one small pond in Edinburgh, Scotland.[54,55]

It should also be stressed that genes at only one locus specify the structure of each individual i-antigen. That this was true was shown by comparing two preparations of i-antigen 90D of *P. primaurelia*, one from stock 90 and the other from a strain derived from the unrelated stock 60, into which the allele d^{90} had been introduced by repeated backcrossing. The peptide fingerprints of the two samples of the 90D protein were found to be indistinguishable and were presumed to be unaffected by any background genes from stock 60.[38] This finding confirms the statement of Preer[31] (p. 306) that "there is only one gene that encodes each serotype and determines the primary structure of the i-antigen serotype."

In addition to the large numbers of allelic variants that can be recognized by immobilization tests with specific antisera in *P. primaurelia*, certain null alleles like the no-F and no-H types mentioned in *P. tetraurelia* earlier also exist. Such null alleles are occasionally found in nature, like the allele F^o in stock 51 of *P. tetraurelia* (see above), though it is possible that in some cases the apparent presence of some null alleles may have been a consequence of ignorance of the precise environmental conditions necessary for their phenotypic expression, or due to their extreme instability. For example, stock 192 of *P. primaurelia* (see Table 7.2) at first seemed to be incapable of forming any G serotype, but by careful breeding work was shown to contain an allele g^{192}, which was apparently never expressed under normal conditions. It was produced by constructing a hybrid between stocks 192 and 90, utilizing as one parent the very stable type 90G. In this way the heterozygote g^{192}/g^{90} expressing the hybrid serotype 192G/90G was formed, and after autogamy yielded, in addition to the very stable type 90G, for a brief period, the very unstable type 192G, and this could be maintained long enough for a specific antiserum (anti-192G) against it to be prepared. Thus, stock 192 was at first erroneously thought to contain a null allele (g^o at the g locus), but in fact was able to produce the type 192G, by expression of the allele g^{192}.[56]

Null alleles, in addition to occurring in nature, can also be induced by various experimental treatments in the laboratory. Thus, mutations to alleles that are unable to produce any H serotype were obtained by x-ray treatment of stock 169 of *P. tetraurelia*.[35] These were shown by Mendelian analysis to be caused by changes at the same locus as that at which the genes specifying the H serotypes H^{29} and H^{169} were situated. Null mutant alleles of certain stocks can be obtained by x-ray treatment of *P. tetraurelia*, followed by selection in the presence of antibodies against serotype A, the major serotype in stock 51.[57] Null mutants in stock 51 of *P. tetraurelia* were obtained, and one of them, denoted d12, was lacking serotype A. Preer (unpublished) then found that in this stock unable to produce serotype A, serotype B was the most stable serotype. So x-radiation of d12, accompanied by selection in anti-B serum, produced nulls for serotype B. One of these, d12.141, could produce neither serotype A nor serotype B. In d12.141, serotype D was the dominant serotype, and radiation of this strain produced mutants null for A, B, and D, in which serotype C was dominant. At this point the process was terminated, but since viability was high, it might have been continued. These double and triple mutants were used extensively in studies investigating the expression of the *A* and *B* genes, as well as in identifying the active gene among the isogenes found at the *D* locus (see below).

In addition to gene mutations induced by x-rays there are also null mutations that are cytoplasmically inherited, appearing after x-ray treatments and screens. They were found to show a remarkable non-Mendelian type of inheritance. The first such case was found by Epstein and Forney.[57] The mutant was called d48 and failed to express antigen 51A. It was shown by molecular means that although the A^{51} gene is present in the micronucleus, in a cross of d48 by 51 it fails to be transmitted into the new macronuclear anlage in cytoplasm derived from parental d48. Thus d48, while having the *A* gene in the micronucleus, lacks the gene in the macronucleus. This results in an apparent cytoplasmic transmission. It should be noted that this is the second kind of epigenesis found in the serotype system, the first being turning off or on the action of serotype loci, as noted above. The epigenetic behavior of d48 is thought to be a consequence of the complex processes whereby macronuclei are differentiated after sexual reorganization in the organism. While this feature of d48 is not produced by activation or inactivation of genes, it is a mechanism that determines whether a gene will be deleted when a newly forming macronucleus is produced. Although it produces a cytoplasmic pattern of inheritance, it could also be considered as a transmission of traits from the old to the new macronucleus. As we will see, its discovery proves to be a most important case. Its epigenetic mechanism is clearly different from the mechanism whereby serotype expression of different loci can be stabilized (mutual exclusion), and provides an explanation for myriad aberrant mutants in *Paramecium*. See Chapter 12.

7.3.4 Control of the Cytoplasmic State by the i-Antigen Genes

In the early work of Sonneborn[58] and Sonneborn and LeSuer[27] on *P. tetraurelia*, emphasis was placed on the role of the cytoplasm in control of the serotypes that could be formed, and a similar role was also apparent in the sytem of antigen variation in *P. primaurelia*.[29] See above. However, it was shown that this cytoplasmic controlling property was in its turn controlled by the antigen-determining genes in the nucleus. This phenomenon is illustrated by the following example.

In *P. primaurelia* a number of stocks collected from different geographic regions were found to develop antigens that were identified as types 60G or 90G by their reactions to the appropriate antisera (anti-60G or anti-90G). Although these stocks were able to produce a G serotype at 25°C, and to transform to a D type at higher temperatures, it was invariably found that those containing the allele g^{60} performed this change (G to D) more rapidly, and at a lower temperature than those containing the allele g^{90}. Moreover, this change was controlled, at least in part, by the gene g^{60} itself, since after making a series of backcrosses by which the allele g^{60} was introduced into a genome containing predominantly genes from stock 90, it was found that cells containing the allele g^{60} combined with a background of other genes from stock 90 were nevertheless able to transform from G to D more rapidly than those containing the allele g^{90} derived from stock 90 itself.[56,59] Capdeville[60] found similar behavior with the alleles g^{156} and g^{168}.[29,42] By experiments of this type, evidence was obtained showing that the antigen-determining genes themselves, as well as other genes, affect the stability of the cytoplasmic (epigenetic) characteristics of the cells at particular temperatures.

The example of the very unstable type 192G of *P. primaurelia*, described above, also can be taken as evidence for the influence of the i-antigen alleles on the cytoplasmic (epigenetic) states of the cells. In *P. tetraurelia* similar results were obtained by Sonneborn et al.[61] and Austin.[47] These authors found that the stability and switching pattern of particular serotypes in stocks 32 and 51 depended partly on which antigen-determining allele was present and partly on other genes, but not on the cytoplasm. It is indeed remarkable and, as it might seem, paradoxical that at least some variations in the stability of the cytoplasm are not transmitted from one generation to the next through the cytoplasm itself, but are somehow controlled by nuclear genes.

In *P. primaurelia*, in spite of the fact that the two types 60G and 90G in homozygous stocks differ markedly in stability, F1 hybrids between these two stocks show no difference in stability of the cytoplasmic state controlling the expression of the two alleles, no matter whether the cytoplasm is derived from the 90G parent or from the 60G parent.[56] Thus, though the control of expression of different antigen-determining genes is determined by the cytoplasm, variations in stability of this cytoplasmic control are determined by nuclear genes.

Caron and Meyer[62] explain this 60 × 90 cross by stating, pp. 28–29, "The expression of a particular antigen gene is thought to produce a specific cytoplasmic factor, which in turn activates or stabilizes the expression of the same antigen gene." In other words, a positive feedback mechanism that is initiated by a gene is postulated. The "cytoplasmic state" proposed by Beale[29] presumably represents a complex of specific factors initiated by many genes that control the expression of the various antigen genes that are found in a given cell.

7.3.5 HETEROZYGOTES AND DOMINANCE

As shown above in the description of crosses between different stocks of *P. primaurelia*, the phenotypes of F1s obtained by crossing two different allelic antigenic types often react as if a mixture of the two parental antigens were present, for individual antibody molecules were found that reacted with both antigens. This evidence suggested to Jones[63] that two polypeptides are present in heterozygotes. Thus, antisera prepared from the heterozygotes g^{60}/g^{90} were found to contain antibody molecules reacting against both parental homozygous antigens, 60G and 90G.[29] Dominance, also known as allelic exclusion (see below), is in these instances absent. Likewise, in *P. tetraurelia*, Dippell[64] showed that immobilization reactions characteristic of both of the two parents appeared in heterozygotes bearing the alleles A^{29} and A^{51}.

Fluctuation in dominance relations has, however, been observed in heterozygotes involving two alleles in *P. primaurelia*. Capdeville[65,60] studied crosses between paramecia of types 156G and 168G and found that the majority of the heterozygotes g^{156}/g^{168} showed the G serotype characteristic of 156G only, while a small minority showed a mixture of both 156G and 168G antigens. A similar result was found using heterozygotes obtained by crossing a stock that, though expressing the allele G^{156}, had been made isogenic with stock 168 by repeated backcrossing. This result was interpreted by Capdeville to mean that the presence or absence of dominance, or, as it was called by her, allelic exclusion, was a hereditary property of the *G* alleles themselves. It was also found that when heterozygotes containing the two alleles G^{156} and G^{168} were transferred to a higher temperature, resulting in a switch to the D serotype, and subsequently returned to a lower temperature at which the G types were again expressed, the mixture of the two types 156G and 168G was again usually found. Therefore, Capdeville et al.[66] confirmed that variation in allelic exclusion (dominance) was not controlled by discrete regulatory genes dispersed throughout the genome, or by cytoplasmic factors, but by the two alleles themselves. However, the variability in phenotypic expression of the g^{156}/g^{168} heterozygotes remains unexplained.

This concludes the account of the basic genetics of serotypes of *Paramecium*. The general conclusion is that genes determine antigen structure and the structure of an antigen is responsible for specificity and the ability of a serotype to be expressed. The actual choice of which gene is to be expressed is controlled by environmental factors, such as temperature, and a system of inheritance that is cytoplasmic in nature. We know little about the mechanism of this system of inheritance, but as we will see, studies on the chemistry of the antigens provide further insight.

One notes that four kinds of epigenetic states have now been encountered: kappa, mating type, serotype, and a kind represented by the mutant d48. We know much more about the d48 system, and we will revisit all the systems in Chapter 12.

7.4 CHEMISTRY AND MOLECULAR BIOLOGY OF THE I-ANTIGENS

7.4.1 ISOLATION OF THE I-ANTIGENS

Chemical study of the i-antigens of *P. tetraurelia* was started by Preer.[24,67-69] It should be stated at the outset that such work has given clear proof that the genetics of the i-antigens is consistent with the Watson-Crick hypothesis of gene functioning by DNA-controlled elements. This raises the question of how *Paramecium* managed to develop epigenetic systems for the control of characters, in addition to the conventional DNA-controlled system of genetic determination. First, however, we will examine the conventional mechanisms.

Preer extracted the contents of cells of *P. tetraurelia* exhibiting particular i-antigens with dilute salt-alcohol, a procedure that does not lyse the cells but breaks off the cilia at their bases and solubilizes the i-antigens and releases them into the medium. These substances were then found to be stable at a very low pH. To purify them, they were first exposed to the very low pH of 2.0, followed by fractional precipitation with ammonium sulfate. They were then shown to be proteins having a molecular weight of 250,000–300,000[69] with a small (2–3%) carbohydrate content.[70] Later it was found that the i-antigens were anchored in the plasma membrane by means of a glycophospholipid.[12,71] These proteins are described below.

The i-antigen proteins extracted from different serotypes yielded very large molecules, differing in size, and with markedly different peptide maps. They differed in their amino acid contents and were not simply different conformations of similar polypeptides.[72–75] Amino acid analyses also showed that the antigens contained high numbers of cysteine residues, involved in disulfide bridges. The amino acid compositions of a number of i-antigens are shown in Table 7.3.[76]

TABLE 7.3
Amino Acid Compositions of i-Antigens

Amino acid	Serotypes					
	51A	51B	51D	90G	178D	90G
Lysine	136	108	197	207	191	138
Histidine	18	17	9	11	9	21
Arginine	35	42	33	35	39	35
Aspartic acid	345	296	334	289	290	271
Threonine	481	417	360	297	313	369
Serine	291	252	193	185	176	169
Glutamic acid	179	134	156	154	176	169
Proline	55	55	70	60	50	46
Glycine	201	174	206	246	224	231
Alanine	354	321	293	237	212	306
Half cystine	328	277	280	252	244	253
Valine	135	108	108	93	82	105
Methionine	15	8	6	7	9	12
Isoleucine	73	67	66	59	60	58
Leucine	113	100	93	104	105	115
Tyrosine	80	62	91	98	98	76
Phenylalanine	45	50	64	70	59	37
Tryptophan	44	33	26	16	16	26
Aspartic + glutamic	306	229	188	140	128	146
Total	*2928*	2521	2585	2420	2338	2406

Reprinted with permission from Reisner, A. H., Rowe, J., and Sleigh, R., *Biochim. Biophys. Acta*, vol. 188, p. 105, Table 19, Copyright 1969 American Chemical Society.

Determination of i-Antigens

According to Jones,[74] the 90D and 60D antigens resemble each other in amino acid composition more closely than they do the G antigens. The type referred to here as 60D was denoted 178D by Jones,[74] because stock 178 was used as the source of this serotype. See Table 7.2.

In the early chemical work on the *Paramecium* i-antigens, it was mistakenly believed that the antigenic proteins consisted of several subunits.[73,74,77] However, Reisner et al.,[78] who used a method of isolating that omitted the use of mercaptoethanol, found no subunits. This finding was consistent with their being determined by single genes since more than one polypeptide would have required more than one gene, and it was early shown that only single genes were involved. See Jones and Beale.[38]

The subunit confusion was cleared up by Hansma,[78] who established that early preparations of extracted *Paramecium* i-antigens had been contaminated by a protease that was specifically activated by mercaptoethenol, and that this protease caused splitting of the antigen protein into polypeptides. The presence of this protease was later confirmed by Davis and Steers,[79] who were able to separate the protease from the i-antigens by ion exchange chromatography. It was then realized that the molecular basis of the i-antigens consisted not of subunits, but of single long polypeptide chains containing about 2,500 amino acids. Boiling an aqueous solution containing the antigens, or even heating them to temperatures below 100°C, however, does not break the polypeptide chain, but did destroy or reduce the ability of the antigens to unite with specific antibodies, the temperature of inactivation depending on the serotype (Beale, unpublished).

7.4.2 THE mRNAs FOR THE I-ANTIGEN GENES

J. Preer realized that since the i-antigens consisted of exceedingly large polypeptides and were very abundant, it should be possible to see them on gels.[80] This proved possible, and when extracts of paramecia were run on gels, large polypeptides were found on these gels that matched precisely the sizes expected for the i-antigens. There were only a few extraneous molecules in their size range, and the very high concentrations of the i-antigens made them easy to identify. Preliminary estimates of their size were given by Preer et al.[36] and later revised.[37] See Table 7.4. Moreover, J. Preer realized that if the proteins were so obvious on gels, then the mRNAs for the antigens should be equally obvious. So RNA was isolated from different serotypes, run through an oligo-dT column to isolate mRNAs, and electrophoresed on a methyl mercury gel for separating RNAs by size. The correlation between the sizes and intensities of the bands of proteins from different serotypes and their corresponding RNAs left no doubt about the identity of the mRNAs.[36] Once the mRNA for a given serotype had been isolated and labeled with ^{32}P, it could be used to probe a DNA library. Thus, it was easy to find the gene.[37]

7.4.3 SEQUENCING OF THE I-ANTIGEN GENES

The initial gene isolated was 51A. Soon the search for additional genes was begun, using essentially the same technique described above. Verification that an isolated clone was indeed the gene was made by probing a collection of whole-cell RNAs from cells of different serotypes with the putative gene that had been made

TABLE 7.4
Mean Sizes of the i-Antigens of Stock 51 Estimated from Gels

Serotype	Molecular Weight
51A	301,500
51B	259,000
51C	257,700
51D	279,500
51E	275,400
51G	308,300
51H	284,000
51I	250,800
51J	283,000
51N	275,000
51Q	256,300

Modified from Forney, J.D., et al., *Mol. Cell. Biol.*, vol. 3, p. 469, Table 1, 1983. With permission from the American Society for Microbiology.

radioactive. A positive test with one of the RNAs and the DNA of the library clone being tested and a negative test with all the other serotypes sufficed to prove that the right gene had been isolated.

An alternative method was used to isolate the G antigen gene of *P. primaurelia*.[81] Meyer et al. started with the mRNA for a given antigen and made a cDNA copy. The macronuclear library was probed with this cDNA and likely clones were isolated. They also used expression of fragments (taking into account the deviant code of *Paramecium*) in *E. coli* to verify that the isolated gene was the correct one.

The following i-antigen-determining genes have now been sequenced. See Table 7.5. References for the table are Prat et al.,[82] Prat,[83] Bourgan-Guglielmetti and Caron,[84] Scott et al.,[85] Preer et al.,[86] Breuer et al.,[87] and Godiska.[88] After the DNA

TABLE 7.5
List of i-Antigens Sequenced

Antigen	Base Pairs	References
P. primaurelia		
156G	8,548	Prat, et al.,[82] Pratt[83]
168G	8,511	Prat[83]
156D	7,632	Bourgain-Guglielmetti and Caron[84]
P. tetraurelia		
51A	8,151	Forney et al.,[37] Preer, unpublished
51B	7,182	Scott, Leeck, and Forney[85]
51C	6,699	Preer, et al.[86]
51D	7,599	Breuer et al.[87]
51H	7,800	Godiska[88]

sequences of various i-antigen genes had been determined, the amino acid sequences of the corresponding proteins were identified using the special genetic code that has been established for *Paramecium*, as described below.

7.4.4 THE UNUSUAL GENETIC CODE OF CILIATES

Study of the DNA sequence of the i-antigens showed that *Paramecium* makes use of a modified genetic code. At the time this discovery was first made it represented, aside from mitochondria, the first case of a deviation from the universal genetic code in any organism. This finding was announced simultaneously by Caron and Meyer,[89] Preer et al.,[90] and Helftenbein,[91] at Cold Spring Harbor at a Gordon conference on ciliate genetics in 1984. It was confirmed later by Horowitz and Gorovsky,[92] working with *Tetrahymena*. They found that *Paramecium*, as contrasted with many other organisms, makes use of the codons UAA and UAG to code for glutamine, instead of functioning as stop codons, as they do in organisms that use the universal genetic code. The first indications of the unusual coding system in *Paramecium* came from the finding that the i-antigen mRNAs in these organisms were not well translated in in vitro translation systems. This was true for the mRNAs from *Paramecium* (and also *Tetrahymena*). In the case of *Paramecium*, a high percentage of the polypeptides synthesized in vitro in the rabbit reticulocyte translation system had a molecular weight of less than 10,000 daltons.[31] Similar problems were encountered in attempts to obtain synthesis of *Paramecium* polypeptides in amphibian oocytes or in *E. coli* expression vectors.[89]

Second, large numbers of UAA and UAG codons were apparently found scattered about in the coding sequences of the i-antigen genes. It was therefore clear that these triplets could not be functioning in the conventional way as stop codons, but were coding for one or more amino acids. To explore likely amino acid assignments for UAA and UAG codons in *Paramecium*, Preer et al.[86] compared the frequencies of amino acids predicted by the observed groups of nucleotides for certain regions of the A^{51} gene in *P. tetraurelia* with the numbers expected from the overall amino acid composition of antigen 51A. It was then found that a best fit was obtained if it was assumed that glutamine was encoded by UAA and UAG. This was deduced from the Chi square value, which measures the magnitude of the discrepancy between the theoretically predicted amino acid content based on codons in the mRNA and the actual amino acid content of the antigenic protein, which was known from chemical analysis as described above. Similar results were obtained by Caron and Meyer[89] with the G^{156} gene of *P. primaurelia*, and by Breuer et al.[87] with the D^{51} gene of *P. tetraurelia*. Only a single stop codon, TGA, occurs in *Paramecium*, and is usually found in clusters of two or three at the end of the coding sequences of the i-antigen genes.

The above-described conclusions concerning the unusual coding system of *Paramecium* was confirmed by reports for some other ciliates in which the triplets UAA and UAG were also found to code for glutamine. It should be mentioned, however, that not all ciliates have the same coding system as *Paramecium* or *Tetrahymena*.[93] For further discussion of abnormal coding systems in ciliates, see Caron,[94] Tourancheau et al.,[95] and Liang et al.[96]

7.4.5 THE I-ANTIGEN GENE SEQUENCES OF *PARAMECIUM*

7.4.5.1 Periodicity

One of the most striking features of the long sequences of amino acids in the i-antigen proteins is the great number of repeats and the periodicity in the spacing of the cysteine residues. This periodicity was discovered by Pratt et al.[82] and Pratt,[83] who analyzed the sequences of nucleotides in the genes that control serotypes 156G and 168G of *P. primaurelia*. Cysteine periodicity was also found in 51A and 51C of *P. tetraurelia*.[97] We will now show how a table of groups of codons, each group beginning with a cysteine, can be built up and how it illustrates the nonrandom arrangement of periodicities. See Table 7.6.

In order to visualize these repeats, consider that the whole 8,000 or so bases of the gene are laid out in a long array like a sentence, left to right. Start at the left of the array and move along to the right until you come to a cysteine codon. This codon and the codons to the right of it, until you come to a second cysteine, are the first group. Start your table of groups and write down the first group of codons, starting with the first cysteine. Now starting with the second cysteine again, continue to the right until you come to a third cysteine, and consider the second cysteine and the codons to its right a second group. Write this group down to the right of the first group with a space or two separating the two groups. Now continue the process until eight groups have been found and written down in your table of groups. These eight groups constitute the first period. Continue now and produce a new period beneath the first one. Repeat the whole process until a second period of eight groups has been found. If the process is continued, a total of 37 cysteine periods will be found. They will constitute the whole gene, and your new table of codons will be complete. Aside from a few discrepancies, these constitute the periodicities.

Examination of your table of codons will reveal that columns of the table will consist of units of differing lengths, but that each column will have units of a unique length. Thus, column 1 typically has a relatively large number of codons per unit, about eleven, while column 8 has a relatively low number of units, about five. This decidedly nonrandom distribution is interpreted to mean that this very large protein has been built up in the course of evolution by duplication of units.

It should also be noted that the 100 or so residues adjacent to the N terminus, and the 150 or so residues adjacent to the C terminus, which occur at the ends of most i-antigen proteins, do not display the cysteine periodicity and are not included in the 37 numbered periods.[83,87,97]

Although the periodicity primarily involves the spacing of the cysteine residues, some other amino acids, such as tryptophan, are also positioned with regularity.[83] Thus, the cysteine and tryptophan residues present a regular periodic distribution in the amino acid sequences. The histidine and methionine residues also coincide nearly perfectly in their positions in the 156G and 168G sequences. Moreover, in all four antigenic types, 156G, 168G, 51A, and 51C, period 10 starts with an exceptionally long group, containing twenty-two to twenty-four amino acids, respectively.[97]

To sum up, there is a cysteine-based periodicity in all the i-antigen proteins that have been studied, but the details vary to some extent in different serotypes, and

TABLE 7.6
Cysteine Periodicity, i-Antigen 156G

1	2	3	4	5	6	7	8
MNNKFIIFSL LLALVASQTY SLTSCTCAQL LSEGDCIKNV SLGCSWDTTK KTCGVSTTPV TPTVTYAAYC DTFAETDCPK AKPCTDCGNY AACAWVESKC TFFTG							
CTPFAKTLDSE	CQAISNR	CITDGTH	CVEVDA	CSTYKKQLP	CVKNAAGSL	CYWDTTNNT	CVDANT
CDKLPATFATDKD	CRDVIST	CTTKTGGG	CVDSGNN	CSDQTLEIQ	CVWNKLKTTS	CYWDGAA	CKDRI
CDNAPTSLTTDDA	CKTFRTDGT	CTTKANGG	CVTRTT	CAAATIGAS	CIKNSSGGD	CYWTGTA	CVDKA
CANTPTTIATNSA	CAGFVTG	CITKSGGG	CVVNGA	CSVANVGAA	CVKNPSNFD	CIWDTT	CKEKT
CAMAPTTNNTHDL	CTSYLST	CTVKSGGG					CQNRT
CANAPTTMTNDA	CEAYFTGNN	CITKSGGG	CVTNTT	CAAITLEAA	CVKNSSGST	CFWDTASSS	CKDKT
CVNATATNTTHDL	CQPFLNT	CTVNSTSAG	CVEKT	CENSLZLAI	CVKDTSSRA	CIWKGK	CYKKQ
CVLASSATTTHAD	CQTYHST	CTLSNSGTG	CVPLPLK	CEAITIEAA	CNLKANGQP	CGWNGSQ	CIDKA
CSTASKFFTTSQ	CTGHIFT	CVAMMPVTVNGLTIQG	CQDLPTT	CARRKSSEN	CEITRVGFPT	CLWVSFSTS	CVEKS
CATASTVGTTGALSAGGSTFSG	CQTYLNT	CISNNTADG	CIAKPFF	CSSLVSSN	CRDGSKASGD	CYWNGSS	CVDKT
CANIIQTTHNS	CNTTSNQ	CTVNNGGTA	CQTLATA	CTSYSTGEN	CKFTSTNKN	CVWTGLA	CRNAT
CADATDTTAYDSDTE	CLAYPTTSET	CTVVYKVGAQG	CVSKSAN	CSDYMTSAQ	CHKTLNLTANDD	CKWIVDR	CYALSSFATGA
CTTFKGTKTM	CEGYRAG	CTNTVGAASSAS					CTLD
CTLKTGSGLTFAD	CGALDST	CSVKKDGTG	CIAIQST	CAGYGSTAAN	CFRSSASGTAGY	CAMNTN	CQSVTSAAE
CAFVTGLTGLDHSK	CQLYHSS	CTSLKDGTG	CQEYKTT	CSGYAATMN	CATSGQGK	CFFDVE	CLRFSN
CASITGTGLTTAI	CGTYDAG	CVANVNGTA	CQEKLAT	CDLYLTQNS	CSTSAAAATADK	CAWSGTA	CLAVTTVGTH
CAYVTGTGLTDLI	CAAYNAN	CTANKAGTA	CQEKKAT	CNLYTTEAT	CSTSAAAATADK	CAWSGAA	CLAVTTVATE
CAYVTGLTGLTDLI	CAAYNAN	CTANKAGTA	CQEKKAT	CNLYTTEAT	CSTSAAAATADK	CAWSGAA	CLAVTTVATE
CAYVTGTGLTNAI	CAAYNAN	CTANKAGTA	CQEKKAT	CNLYTTEAT	CSTSAAAATADK	CAWSGAA	CLAVTTVATE
CAYVTGLTGLTNAI	CATYNAN	CTANKAGTA	CQEKKAT	CNLYTTEAT	CSTSAAAATADK	CAWSGAA	CLAVTTVATE
CAYVTGLTGLTKAI	CATYNAG	CINLKDGTG	CQEAKAN	CKDYTTSNK	CTAQTTSTLS	CLWIDNS	CYPVPDLN
CSVITGLGFVHAQ	CQAYSTG	CTSVSDGSK	CQDFKST	CEQYPGTTLG	CTKTASPK	CYLQGSA	CITISNVATD
CAKITGSAGTITFEI	CQSYNTG	CSVNRARSA	CVQQQAQ	CSGYTSAMTS	CYKSGAGL	CIASTNTDTA	CVAATAAT
CDAVYLGAGNYSSAN	CNEMKAG	CTNNGTTA					CVAKT
CANAAGITFNHTN	CNSYLNT	CTVNSGNSA	CQTMASK	CADQTQAS	CLYSVEGE	CVVVGTS	CVRKT
CDTAATDATRDDDTE	CSTYQQS	CTVARLGA	CQARAA	CATYKSSLQ	CKFNTSGGK	CFWNPTNKT	CVDLN
CGNIEATTLYDTHNE	CVAVDATLA	CTVRATNGAAAAQG	CMARGA	CASYTIEEQ	CKTNASNGV	CVWNTNANLPAPA	CQDKS
CTSAPTSTTTHND	CYAYYNTATVK	CTVVATPSNSGGNPTLGG	CQQTAA	CSSYIDKEQ	CQINANGDP	CGWNGTQ	CADKS
CATASATADYDDDTK	CRAYITNK	CTVSDSGQG	CVEIPAT	CETMTQKQ	CYNKAGDP	CYWTGTA	CITKV
CDNAPDATATADE	CNTYLAG	CTLNNVK					CKTKV
CEDFAFATDAL	CKQAIST	CTTNGTN	CVTRGT	CFQALSQAG	CVTSSTNQQ	CEWIPAVLNASNVITSPAY	CTIKN
CSTAPITLTSEGS	CAGYFTN	CTTKNGGG	CVTKST	CSAVTIDVA	CTTALNGTV	CAWDSAQNK	CRDKD
CQDFSGTTHAA	CQAQRAG	CTAGAGGK	CARVQN	CEQTSVRAA	CIEGTNGP	CLWIDKYQNTDGTKGA	CFRYTS
CKSLNWNNDS	CKWISNK	CTTNGSN	CVGITL	CSETNTDGG	CVTGYDGA	CIQSVPDLNSSDPKV	CKPYTS
CADAFYTTHSD	CQIASSK	CTTNGTTG	CIALGS	CSSYTVQAG	CYSNDKGTLYTSGVITSTGI	CTWDTTSSS	CRDQS
CADITGTTHAT	CSSQLST	CTSDGTT	CLLKGA	CTSYTTQTA	CTTAVGSDGA	CYWELASATNNNTAK	CRLLT
CADIQNGTATNV	CSVALST	CVSNGTA	CIPKAN	CSTYTSKVA	CNSGGLDGI	CVFTQSTATGAAAGTGT	CALMTA
CTVANNDQTA CQAARDRCSW TAASGTRATA VASKCATHTC ATNGATNGAC TRFLNWDKKT QQVCTLVSGA CTATDPSSFS SNDCFLVSGY							
TYTWNASTSK CGVCTAVVVQ PNTTDNNTNT TDNNTTDSG YILGLSIVLG YLMF							

This article was published in *J. Mol. Biol.*, vol. 189, Prat et al., Nucleotide sequence of the *Paramecium primaurelia* G surface protein, a huge protein with a highly periodic structure, p. 58, Fig. 8. Permission granted by Elsevier, who holds the Copyright, 1986.

although some amino acids other than cysteine also show a certain degree of periodicity, this is less striking than that of the cysteine residues.

7.4.5.2 Tandem Repeats

When the bases within a single gene are compared, they are found to have repeated sequences of around two hundred bases near the center of each molecule. (Here, in considering tandem repeats, we are comparing sequences within single genes, not between alleles or genes, as in the next section.) In some serotype genes, for example, those controlling 156G (see Table 7.6; note this table gives amino acid sequences, not base pairs), and in 168G, i-antigens of *P. primaurelia*, and 51A of *P. tetraurelia*, there is a remarkable regularity within the sequences of each serotype in the composition of the centrally located periods. In 156G and 168G, all periods between 14 and 23 show some degree of similarity, and this is especially marked for the five periods between 17 and 21, which contain seventy-three (in 168G) and seventy-four (in 156G) amino acids. Note that one period contains $3 \times 73 = 219$ base pairs. The central regions of these genes contain series of nearly similar, or in some instances identical, tandem repeats,[83] and this has been shown to apply not only to the sequences in the macronuclear DNA, which are expressed in the phenotype, but also to those in the micronuclei.[98] A similar situation occurs with the 51A i-antigen of *P. tetraurelia*, where the central region consists of a series of five almost identical tandem repeats, each comprising eight cysteine residues and about seventy-five other residues.[97] However, other i-antigen sequences lack these central tandem repeats. They do not occur in 51C,[97] 51D,[87] or 156D,[84] though 156D has a repeated motif at both ends of the sequence.

7.4.5.3 Allelic and Nonallelic Variations

Comparison of the sequences of bases in the DNA of genes for different allelic serotypes was studied first in the *P. primaurelia* genes G^{156} and G^{168}. See Table 7.5. The two antigens produced by these alleles do not show any immunological cross-reactivity by the immobilization test,[99] but their allelism has been confirmed by Mendelian analysis.[100] Both were found to be located in *Bam*H1 restriction fragments near the telomere of a macronuclear chromosome. The complete nucleotide sequences of these two genes have been determined.[82,83] It is of interest to compare the base sequences of these two alleles with each other (not comparing one repeat to another as was done in the preceeding section). They differ in regard to some 487 nucleotide substitutions. Of these, 256 are silent, that is, affect only nucleotides, not amino acids, and 231 involve 194 amino acid changes, some codons undergoing two or three substitutions. Seventy-two percent of the amino acid substitutions, from residue 1030 to residue 1504, are clustered in the central parts of the genes (Prat,[83] p. 527). The amount of variation in the central region of these genes is thus relatively large, and is also noteworthy for being concentrated in the region where the central tandem repeats are found. Nevertheless, in spite of the many differences between these two alleles, it is stated by Prat[83] that the open reading frame of 8115 bp of antigen 168G is highly homologous with the 8145 bp of the allelic protein 156G, and according to Bourgain-Guglielmetti and Caron,[84] the similarity between the two allelic sequences may be considered extremely high (98%), except for the central repeats, where the similarity percentage drops to 60%. Similar results were obtained for the two immunologically similar allelic serotypes 51A and 29A of *P. tetraurelia*, where the similarity is reported as 99.5% over most of the coding sequence but only 77% in the central repeat region.[101]

Comparisons between the amino acid sequences of different (i.e., nonallelic) serotypes sometimes indicate unexpected similarities. For example, the serotypes 51A (from *P. tetraurelia*) and 156G (from *P. primaurelia*) are surprisingly similar in the sequences of their controlling genes A^{51} and G^{156}, which were found to be 80–90% identical in both base and amino acid sequences within one thousand bases of the 3' and 5' ends.[86] In the central portions of these genes the homology is somewhat less, but is always close to 70% or higher. The homology between the genes A^{51} and C^{51} is much weaker, and coding regions of the gene H^{51} also show only weak homology with any of the other genes studied. However, the amino and carboxy terminal regions of A^{51}, C^{51}, and G^{156} can be aligned with few extensive gaps.[97] Further, Breuer et al.[87] (p. 317) state that the D^{51} gene is not homologous with any of the other i-antigen genes (other than the D isogenes), except the d^{156} gene of *P. primaurelia*, which it strikingly resembles, but as these two genes occur in different species, their allelism cannot be established by Mendelian methods. Usually there are many differences between the amino acid compositions of different nonallelic i-antigens, though sometimes, as with the amino acids of the two alleles g^{156} and g^{168}, the molecules are relatively invariant at the two ends. This lack of variation in the distal regions seems to be characteristic of both allelic and nonallelic comparisons.

To sum up this account of variation in the coding sequences of different i-antigen genes, it may be concluded from the incomplete data available that most variation occurs in the central regions, especially in those genes having tandem repeats in those regions. There is less variation in the terminal parts of the coding sequences, and least of all in the surrounding noncoding sequences (as discussed below).

Not much is known about the correlation between variation in nucleotide sequences of i-antigen genes and immunological diversity of the cells. According to Caron and Meyer,[102] during the immobilization test only a small part of the i-antigen molecule is involved in interactions with antibodies. This view is based on the observation of Capdeville[99] that although two antigens such as 168G and 156G may show no cross-reaction when compared by the immobilization test, nevertheless these and all other allelic i-antigens in a series display common antigenic determinants when examined by immuno-diffusion tests. It has been suggested that the i-antigen molecule is composed of two parts, one accessible to antibodies in vivo and the other masked, or consisting of elements that are not exposed to the exterior environment. It may be speculated that the exposed regions are those concentrated in the central and most variable part of the molecule. The repeats in this region differ in different serotype genes.

7.4.5.4 Start and Stop Codons

The genes A^{51}, B^{51}, and D^{51} of *P. tetraurelia* have short (ca. 8 bp) transcribed but untranslated segments immediately upstream of the AUG translation start site.[86,87,103-105] However, no clear promoter or other sequences have been identified within the first hundred bases upstream of the start of translation at AUG. As we will see, a promoter has been found for the A^{51} gene in the region −264 to −211. See Section 7.5.6.

Most i-antigen genes analyzed have two or three UGA stop codons, and the end of translation is taken as the first UGA stop codon. The end of transcription was determined by Godiska[88] using an S1 protection assay and found to be fifty-five bases past the first TGA for serotype 156G and at base 65 for serotype 51H. Although it has not been determined for 51A, sequencing of a DNA complement of mRNA for the end of the A gene reveals that the polyadenylation site begins at base 46. See Table 7.7.

7.4.5.5 Isogenes

In both *P. tetraurelia* and *P. primaurelia* there are DNA sequences that are very similar to those of the previously identified D genes. This peculiarity of the D locus first became evident when mRNA from 51D cells was used to identify genomic clones containing complementary DNA sequences. When D-specific mRNA of *P. tetraurelia* was used as a probe for screening different genomic libraries, several similar but not identical sequences were found.

Schmidt[106] and later Breuer et al.[87] isolated five related genes: 51 alpha, beta, gamma, delta, and epsilon. The sequenced portions of all these genes were highly homologous, and this applies not only to the coding regions (which are generally similar among i-antigen genes), but also to the noncoding regions.

Schmidt was able to identify which of the genes was the one for serotype D by examining the triple 51 deletion mutant, missing the A, B, and D genes, and observing which of the five genes was missing. This turned out to be 51 alpha D.

A similar situation was found at the D locus of *P. primaurelia*,[84] where three very similar genes were identified. Two of these genes were physically close together, but oriented in opposite directions. Bourgain-Guglielmetti and Caron were able to identify which of the genes was transcribed and represented the true gene for serotype D (it turned out to be 51 alpha) by looking at the mRNA produced in serotype D. It is possible that the various isogenes of i-antigen D represent genes for other less frequently expressed serotypes that cross-react with serotype D.

At first it was thought that these genes (all but the alpha forms) were nonfunctional and they were designated pseudogenes. Later it was realized that the genes contained no stop codons and hence were probably fully functional. They were then designated isogenes and thought to be genes for unexpressed serotypes, such as I, J, and so on, or possibly even genes for new or undiscovered i-antigens.

7.5 CONTROL OF EXPRESSION OF I-ANTIGEN GENES

7.5.1 Hypotheses to Explain Control

As was stressed at the beginning of this chapter, one of the major points of interest in the i-antigens is how the cell selects a single i-antigen for expression on the cell surface, and how this expression can be inherited cytoplasmically over many fissions. The first attempt to explain it was the plasmagene theory, but as we have seen, the plasmagene theory was finally rejected on the basis of the work on kappa. The

TABLE 7.7
3' Ends of the Coding Sequences of the i-Antigen Genes

```
P. primaurelia
156 G          TGA TGA GAA TGA TTTTTAAATA TATATAGGGA AAATTTTATA AAATACACAA AAAATCATCT TTCTAATCAT TATTTTTGT
168 G          TGA TGA GAA TGA TTTTTAAATA TATATAGGGA AAATTTTATA AAATACACAA AAAAATCATC TTTCTATTCA TTATTTTTTT
156 D alpha    TGA TTT TGA TTT TCATATTATA TAGGGAAAAT TTTAATATTA CAAATTATTT TTATCTCTCA ATAGAGGGTT TTTTAATAAA
156 D beta     TGA TAT TGA TTT TCATATTATA TAGGGAAAAT ATTAATAATA CAAATTATTT ATCTCTTAAT TAGTAATATT GAGGTGTTTT
156 D gamma    TGA TTT TGA TTT TCATTTATA GGGAAAATAT TAATATTACA AGTTATTTTA TCTCTTAATA GAGGTGTTTT AGTAGATGTT

P. tetraurelia
51 A           TGA TGA GAA TGA TTTTTAAATA TATATAGGGA AAAATTTATT AAATATAAAA AAAATCTTTA TAATTATTAA GATATAGAAA
51 B           TGA TGA GAA TGA TTTTATAAAT TTATATAGGG AAAAATTATC TTAAATTTAA AACAAATATT TCTTATTTTT GGTTATAAAT
51 C           TGA TGA ATT GAT TCATTAAATT TATATAGGGA AAATTTTATT AAATAATCAA AAAGTATTCA TATAATCATT ATATATATCA
51 D alpha     TGA TTT TGA TTT TCATATTATA TATAGGGAAA AATTTATTAA TGAAAAAAAT TAATTTAATC TTTAAATAAG ATTGTTTCAA
51 D gamma 1   TGA TGA TCA TTT ATAATATGTA TTTTCAATTC ATATCATGTA ATTTATTTTB TATCAAATCT CATACTGGTT GAATAAATTA
51 D gamma 2   TGA TTT TTA TTT TCATATTATA TATAGGGAAA AATTATTAAT GAAATAAATA AATGATTGTT TCAATQAATA AATAAATCTA
51 D delta     TGA TTT TGA TTT TCATATTATA TATAGGGAAA ATATTAAACAT TACAATTTAT TTTCATCTCT TAAAACTTCT TATTATTTAA
51 D epsilon   TGA TTT TGA TTT TCATATTATA TATAGGGAAA AATTATTAA TGAAAAAATT AATTAATCT TTAAATAAGA TTGTTTCAAA
51 H           TGA TTT TGA TTT TAAATTTATA TAGGGAAAAT TACATATATT AATAAAAAAA TTCATTTCTT ACAATCAATT TTGATAGAAA
```

Sequences from following sources: 156G Prat, et al., 1986[82]. Godiska, 1987[88]. 168G Pratt, 1990[83]. 156D Bourgain-Guglielmetti and Caron, 1996[84]. 51A Preer, unpublished. 51B Forney, unpublished. 51C Preer et al., 1987[86]. 51D Breuer, et al., 1996[87]. 51H Godiska, 1987[88].

second attempt was by Delbrück.[43] He assumed a system of synthetic steps leading from each antigen gene to the final product, in which one of the products in each line of syntheses inhibited all the other syntheses, but not its own. He noted that in this competition, if one of the sequence of steps obtained a slight advantage, then it would take over and quickly become the major synthetic pathway. Although this theory was cited even before the steps in protein synthesis were known, this theory was still given much later by Monod and Jacob.[44] They noted that Delbrück's model assumes allosteric inhibitions (inhibition by a product molecule at the end of a synthetic sequence of the first step in the synthesis) known to occur in some systems. It appears to us that the chief problem with this model is that it requires the unlikely assumption that there are about a dozen reactions that inhibit all the other serotypes except the major molecule synthesized. Monod and Jacob then go on to give a second model, based on regulators and operons. In this model each i-antigen gene has a regulator and an operator, in which the i-antigen gene has been turned off by a product of the regulator locus, which has combined at the operator locus. This represents a classic regulator-operator induction system, except that the inducer is not the substrate, but a product of the synthesis itself, such as the i-antigen.

This model is self-perpetuating. Moreover, it would not need an explanation of why the other i-antigen loci are turned off, for it is assumed that all loci are off unless they are specifically turned on. It would appear to be adequate to explain the i-antigen system. It would be equivalent to the model proposed by Kimball[107] of self-induction of the i-antigens without such detail as to how the process occurs.

7.5.2 POSITIONAL CONTROL

In another protozoan, *Trypanosoma*, there is a situation superficially similar to that of the i-antigens of *Paramecium*, but in *Trypanosoma* there is evidence that antigenic variation is produced partly as a result of shifts in the chromosomal location of the genes concerned,[108,109] the gene moving in and out of expression or nonexpression regions of the chromosome. In *Paramecium*, however, no chromosomal rearrangements are known that are correlated with changes in the expression of the i-antigen genes. The structures of these genes and their neighboring sequences on the chromosomes are the same whether the genes are expressed or not.[37] Hence, antigenic variation in *Paramecium* is controlled by a different mechanism from that occurring in *Trypanosoma*.

It is perhaps relevant to this question that some, perhaps the majority, of the serotype genes in *Paramecium* are located near to the terminal telomeres of the macronuclear chromosomes. Two of the genes, A^{51}[37] and G^{156}[89] are within 15 kb of the end of a macronuclear chromosome and are oriented with their 3' ends toward the telomeres. On the other hand, the 3' end of the gene H^{51} is not less than 18 kb from the telomere, but there is no evidence that the position of this gene affects the expression of the gene.[88] Whether there are any other genes between the serotype gene in the case of serotype H and the end of the macronuclear chromosome is unknown. In general, it has not been shown that the expression of the i-antigen genes of *Paramecium* is affected by their proximity to the telomeres at the ends of the macronuclear chromosomes; however, a careful study of G^{156}/G^{168} heterozygotes[110] revealed that the amount of mRNA made by a gene is affected by its position relative to the telomere. No satisfactory hypothesis was found to explain this phenomenon.

7.5.3 EARLY OBSERVATIONS ON SEROTYPE SWITCHING

Much of the early work on this subject consisted of efforts merely to examine the nature of the problem. The work began with Sonneborn and LeSuer[27] and Beale[59] on serum-induced changes in *P. tetraurelia*. The change from serotype 51D to 51B after treatment of 51D paramecia with anti-51D serum was studied, at the phenotypic level, by observations on immobilization of living cells by the two antisera anti-51D and anti-51B. During this change the switch (or transformation, as it was formerly called), which occurred after the initial treatment with anti-51D antiserum, involved cessation of formation of one antigen (51D) and commencement of formation of another (51B).

Details of the location of the i-antigens on the cells, and of the process of switching from one serotype to another, have been revealed by studies with labeled antibodies, either using fluorescent substances, which can be observed by ultraviolet light microscopy,[13,14] or by observations on ferritin or immuno-gold preparations by electron microscopy.[12,16,111] Such experiments showed that the i-antigens are substances coating the entire surface of the pellicle and cilia of paramecia. Evidence was also obtained showing that the antigens are synthesized on ribosomes in the cytoplasm, and subsequently transported to the exterior of the cell.[112]

The serotype switch from 168G to 168D in *P. primaurelia*, caused by a rise in temperature, was studied by Sommerville.[113,114] In practically all stocks of *P. primaurelia*, the change from serotypes G to D can be brought about by raising the temperature at which the cells are grown, but the rate at which this change occurs varies greatly according to the genotype and other factors. Usually there is a period after raising the temperature during which the G type continues to be formed, and after a shorter or longer time, there is a relatively sudden switch, lasting for about the time of two fissions, to the D type appropriate to the genotype of the particular stock under study. During the switch itself, both i-antigens (G and D) can be detected in a single cell.

In the experiments of Sommerville, paramecia of type 60G were transferred from 18°C, at which this serotype is stable, to 35°C, when the switch to serotype 168D occurs rapidly. During the time the cells were expected to change their serotype, they were grown in medium containing radioactively labeled ^{35}S bacteria, and synthesis of the G and D i-antigens was followed over a period of about three fissions. At various times samples of cells were harvested and the relative amounts of G and D antigens were assayed by making autoradiographs from immunophoresis gels. It was found that radiolabeled D antigen was detectable in samples taken as soon as 15 minutes after change of temperature. Clearly there was an immediate response to change of temperature affecting expression of the gene D^{168}, but the old type (G) antigen also continued to be synthesized by the gene G^{168} for at least 2 hours after the temperature change.

Earlier it had been thought by Sonneborn, working with *P. tetraurelia*, that the change from serotype 51D to 51B, induced by treatment with immobilizing anti-51D antiserum, was caused by a change in some hypothetical cytoplasmic genetic factor. However, after the role of the nuclear genes in determination of the i-antigens was established, both in *P. primaurelia* and in *P. tetraurelia*, it became clear that the antigenic changes, whether spontaneous or induced, should be interpreted in terms of the calling into action of certain genes, either by something in the cytoplasm or by some effect of the environment. For example, the change from serotype 51D to 51B in *P. tetraurelia* would then be thought to be due to a cessation of control of antigen synthesis by the gene D^{51} and its replacement by the gene B^{51}. Simultaneous activity of the two genes would only continue for a short time, lasting for about two fissions. It should be remembered, however, that the expressed genes controlling the phenotype are located in the macronucleus, which contains several hundred copies of each gene. Hence, not all the copies of a given gene in a single *Paramecium* may be switched on or off simultaneously.

7.5.4 IMPORTANCE OF GENE STRUCTURE

There is evidence that antigen variation in *Paramecium* involves changes in the functioning of genes at the level of formation of mRNA, that is, of transcription. This was first apparent from the finding that *P. tetraurelia* agarose gels containing extracts of cells expressing one serotype show only a single band of mRNA of a size corresponding to that necessary for the synthesis of the very large i-antigen proteins, and the relative sizes of these mRNA molecules are characteristic for each serotype.[36,37] In *P. primaurelia* it has also been shown that i-antigen production is regulated at the level of transcription. This was proved by experiments using cDNAs obtained from mRNA of G and D types of stock 156 of *Paramecium* as probes with total mRNA from the two antigenic types, and showing that mRNA corresponding only to the expressed antigen was present in the cytoplasm of the cell.[110,115]

At this point in the argument it is necessary to discuss the problem of what factors control the expression of the antigen-determining genes. Early studies indirectly supported the idea that the antigen genes themselves play a role in controlling their own expression. As already stated, Sonneborn and LeSuer[27] and Beale[29] showed that the apparent control of the cytoplasm on the expression of the antigen-controlling genes is ultimately traceable to the action of the i-antigen genes themselves. Moreover, while different alleles at a locus usually do not show mutual exclusion, in some cases they do. Thus, Capdeville[60,99] showed that exclusion occurred between the alleles involving g^{156} and g^{168}. She found that exclusion was a property of the antigen alleles themselves, rather than of other genetic factors.

7.5.5 RUN-ON TRANSCRIPTION

Another approach to the control of i-antigen gene expression is by studying the production of mRNA in run-on transcription experiments carried out on *P. tetraurelia*.[116] These experiments used cytoskeletal frameworks, that is, cells that had their cellular membranes removed, but still contained macronuclei with DNA capable of extending synthesis of mRNAs in vitro when provided the proper substrates. Labeled mRNA fragments were synthesized in vitro and were used as probes for hybridization to various cloned DNAs. Transcription of the genes A^{51}, C^{51}, and H^{51} of

P. tetraurelia was studied and labeled mRNAs were obtained from them by run-on transcription using each one of these serotypes. It was shown that the A serotype produced only labeled A mRNA and the C serotype produced only C mRNA, but H mRNA contained not only the expected H mRNA, but also C mRNA. It was concluded that the genes A^{51} and H^{51} were regulated at the level of transcription, failing to produce heterologous transcripts. It was also concluded that gene C^{51}, which appeared to be active in the serotye H cells, was not inhibited until after transcription. This conclusion was supported by their demonstration that while serotype A and C cells contained only homologous RNA, serotype H also contained C RNA. Moreover, a look at the three proteins showed that each of the three serotypes produced only its homologous protein at the surface of the cell. There were no traces of the C protein that could be extracted from serotype H. Thus, it was concluded that in serotype H there are two methods of regulation. In one case (i-antigens A and B) it is at the level of transcription, for no message is made. In the other case, it is at the level of translation (i-antigen C), for i-antigen messenger is made, but no i-antigen is synthesized.

Leeck and Forney,[104] in a similar run-on experiment, showed that a culture expressing the antigen 51A at 27°C transcribed the gene A^{51} but not the gene B^{51}. Likewise, cells expressing the antigen 51B transcribed the gene B^{51} but not A^{51}. Therefore, expression of both genes A^{51} and B^{51} was controlled at the level of transcription.

7.5.6 Promoters of i-Antigen Genes

Several investigators have used the technique of microinjection to search for promoters. We diverge a moment to decribe this technique. It is remarkable that injections into the macronuclei of *Paramecium* of pieces of DNA, whether derived from *Paramecium* itself or from some quite different source (e.g., plasmids, bacteriophage, etc.), result in replication of the introduced fragments of DNA and, in the case of injected *Paramecium* DNA, expression of their contained genes. The DNA is normally maintained for the lifetime of the recipient macronucleus, that is, until the next autogamy when the macronuclei disintegrate and are reformed.[117–120] Replication can occur either as linear telomerized molecules or as chromosomally integrated molecules.[121] See also Haynes et al.,[122] who showed expression of many injected genes.

With the aid of this procedure, Leeck and Forney[104] were able to show that the region necessary for the A^{51} gene (the promoter) lay in the 124 bases between –273 and –150. This result was not quite as precise as that obtained by Martin et al.[103] in a paper published a few months before the Leeck and Forney paper. Using a different set of deletions, Martin et al. found that there were fifty-four essential bases in the region –264 to –211. There are only four published sequences that extend far enough upstream for comparison: 156G, 168G, 51A, and 51H. These sequences are given in Table 7.8. It is noted that in the fifty-four-base region containing the promoter of the A^{51} gene there is the ten-base-pair sequence, TATATAAGTT, in both the A^{51} and H^{51} genes. However, this sequence is not found in the B gene or in the 156G or 168G genes. A common promoter in higher organisms is TATATAAA, the TATA box.

Thai and Forney[123] were able to show by deletion analysis that there are still other regions of the A^{51} gene that are necessary for expression of serotype A. They found

TABLE 7.8
5′ Ends of i-Antigen Genes

```
        -270                                                                                                                                              -136
156 G   GGGAAAAAAA ATAACCTATT TAATTTGATA CCAAGGTCTT AATGCTTCCA ATCAGGCTCT TCAACATTTT GAAGCTATCT TCACACACCC ACCTTCCCTT ATAACATTAA AAACAAATAA TAAATTAATT AATGA
168 G   TATACTTGTG AATATAATAT AAGCAATTTA ATCTAATACC AAGGTTCTTA TGGTTCCAAT CTGAGTTTTC TAAATTTTGA AGTTATCTCC ATCTTCCCTT ATAACATTAA ATACAATATA TAAATTAATA AATGA
51 A    GCAATATTTA TATAAGTTTT GAATGCTGTA TAGAAGAGTG AAGATTAAAT ATTAGCTTTT CCAATTCTAC CCTTAGAATT GAACTAGTTT CAAAAATAAA TGGTAATTAT ATTATTAATA TGAATTAAAG ATTTT
51 B    ACAATAACTT AAAATTTTGA AAAATCAAAA AGTAGAAAT GAAGAACACA TTCAGATTGT CTTCCTTCAA TTTCATCACT TTAAGCCTTC CATACTAATC CAATTTACAT TATTTTAAAT ATTAACATAA ATTGA
51 H    AATAAATATA TAAGTTCATA ATATCATCAA ATAGAATACA AACACTCTGT TAGAAGCCAT CTGTGTTTTC ATTCTCAAGT TAGTCTAGAT AATTATATTA AAATTAATGA TGATTATGAT GGAGG

        -135                                                                                                                                              -1
156 G   TGATG ATTAATATTA TATTTAAATT CTGATTTACA AACAAGAAAA TATTATTAGA TCATAAATTA AATTAAAATA TTTAATTTGA AAAGCGATA TCCAGTTAT TTTGAATTTT AATACTTTTA ATG
168 G   CGACG ATTAATATCA TATTTAAATT CTGATTTACA AACAAGAAAA TATTATTAGA TCATTAATTA AATTAAAATA TTTAATTTGA AAAGCGATA TCCAGTTTA TTTGAATTTT AATACTTTTA ATG
51 A    GATAT GAATTAATGA TTAATGTGAA TTTTGAGGCT NAACTTATTT AAACAAAATG AGGTAATTGA AAAAAAAAAA AAATTAACA AACTAACTAT TTAAAAATTC ATTCGATTCA AATTTAATTT TAATACTTTA ATG
51 B    GGTCA ATTATGTTGA AATTTTAATT CTGCAAAATA AACAATTACT AAAAACTGAT GTTTAATTAA AATTAAATTT TATTCATAAT TTAGCTTGAT TAGAATGTAA ACCTAAAGTT AATACTTTTA ATG
51 H    ATTAT TAGATAAAAT TAAATTTCCT AAATTATATA TTTAATAATA TGATATAAGA TAAAAATTAA AAAAATTTAA TTTTTCATAT ATTTTAAATT GAAAAAGTAG ATCCTTAAAT ATTTAATTA CTAAATAAAA ATG
```

Sequences from: 156G Prat et al., 1986[82]. 168G Prat, 1990[83]. 51A Preer, unpublished. 51B Forney, unpublished. 51H Godiska, 1987[88].

that deletion of a small section in the C-terminal portion of the gene was necessary for expression.

7.5.7 THE MUTUAL EXCLUSION TEST

As was long ago shown by Sonneborn and LeSuer[27] and Sonneborn,[32] the two serotypes 51A and 51B show mutual exclusion, for each can be maintained indefinitely in separate cultures under the same environmental conditions, by growth at one fission per day at 27°C. Under these conditions, the synthesis of i-antigen B is excluded in serotype A-expressing animals, while the synthesis of i-antigen A is excluded in serotype B-expressing animals.

By injecting an upstream region of the gene A^{51} into macronuclei of cells of the null mutant d12, which lacks the A^{51} gene, Martin and colleagues[103] showed that this upstream region contained the promoter that was necessary for expression of this gene, but the same authors also concluded that this upstream region was not involved in the control of mutual exclusion of other serotypes by the A^{51} gene, or of the responses of serotype 51A to environmental (e.g., temperature) changes. This conclusion was based on the finding that in spite of the marked reduction of expression of the gene A^{51} as a consequence of loss of certain upstream regions between positions -211 and -264, mutual exclusion could still occur. This was because, in spite of lacking this upstream region, some cells did nevertheless express the gene A^{51}, although to a reduced extent. These cells still showed the usual exclusion of the 51B and other i-antigens at 27°C. Such rare cells of type 51A showed the same responses as wild-type stock 51 cells to temperature and treatment with homologous antiserum, resulting in formation of type 51B at 19°C. In no cases were cells produced that were capable of producing both A and B antigens on the cell surface at the same time. So, while the promoter is necessary for gene expression, it does not control mutual exclusion.

Leeck and Forney[105] were able to demonstrate that the portion of the gene concerned with mutual exclusion lies not in the upstream region where the promoter lies, but farther along in the transcribed region of the gene. In order to do this, they used the promoter of the B gene attached to sections of the A^{51} gene. They found that if they made such hybrids, the B portion functioned normally like a promoter for the A^{51} gene, and could largely be ignored in viewing the results of the experiments, the element of importance being the extent of the region of the A^{51} gene included in the construct. Such hybrid molecules along with a whole functional B^{51} gene were injected into a double mutant in which both the A^{51} and B^{51} genes had been deleted, and the character of the cells then ascertained.

First is the element that controls mutual exclusion found to the right of +1. They injected the whole transcribed A^{51} gene to the right of +1 (with appropriate B primer on the left of the construct) and found that either serotype A or serotype B is expressed in the recipient, but never both on the same cell. So the region that controls mutual exclusion must be to the right of +1.

Next, they moved the break site down into the transcribed portion of the gene to +885. They found that portions of the B^{51} gene on the left, while containing a small section of B, were still able to work with the A^{51} gene and produce A proteins.

Moreover, cells injected with this construct showed both A and B antibody sites on a single cell. The transcribed region between +1 and +885 contains a region necessary for mutual exclusion, and it was deleted in this experiment.

They carried out additional experiments that confirmed this conclusion. Thus, the sites on the gene that are necessary for mutual exclusion are within a few hundred bases downstream of the start of translation. This result fits nicely with Kimball's hypothesis that the immobilization antigens control their own synthesis, and with Capdeville's conclusion based on genetic studies that mutual exclusion is a property of the antigens.

7.6 CONCLUSIONS

It has been important in understanding the i-antigens to understand that some regions of the gene are involved in simply turning the gene on and off. One of these regions is the promoter and is found in an upstream noncoding region as expected. There is another region in the downstream portion of the gene that is also necessary for expression. But the finding that in the immobilization antigens there is still another region that can also turn the gene off and on is of major importance. This is the region that is concerned with mutual exclusion and has been localized to bases +1 to +885. Of course, knowing where something works does not tell us how it works. Now we need to find the specific enzymes and factors that enable binding to this chromosomal region before a satisfactory explanation can be obtained. When we learn more about these factors and how they work, we will learn how the different serotypes can be expressed in a hereditary manner, that is, mutual exclusion. Possibly a factor that binds to this site will even behave in the manner postulated by Delbrück[43] or Monod and Jacob,[44] or, perhaps more likely, an entirely new mechanism will be found.

REFERENCES

1. Koizumi, S. 1966. Serotypes and immobilization antigens in *Paramecium caudatum*. *J. Protozool.* 13:73.
2. Steers, E. Jr. and Barnett, A. 1982. Isolation and characterization of an immobilization antigen from *Paramecium multimicronucleatum*. *Comp. Biochem. Physiol.* 71B:217.
3. Doerder, F. P. 1973. Regulatory serotype mutations in *Tetrahymena pyriformis*, syngen 1. *Genetics* 74:81.
4. Sonneborn, T. M. 1975. *Tetrahymena pyriformis*. In *Handbook of genetics*, ed. R. King, chap. 19. Vol. 2. New York: Plenum Press.
5. Williams, N. E., Doerder, F. P., and Ron, A. 1985. Expression of a cell surface immobilization antigen during serotype transformation in *Tetrahymena thermophila*. *Mol. Cell Biol.* 5:1925.
6. Doerder, F. P. and Berkowitz, M. S. 1986. Purification and partial characterization of the *H* immobilization antigens of *Tetrahymena thermophila*. *J. Protozool.* 33:204.
7. Love, H. D., et al. 1988. mRNA stability plays a major role in regulating the temperature-specific expression of a *Tetrahymena thermophila* surface protein. *Mol. Cell. Biol.* 8:427.
8. Rössle, R. 1905. Spezifische sera gegen Infusorien. *Arch. Hyg.* (Berlin) 54:1.
9. Finger, I. 1974. Surface antigens of *Paramecium aurelia*. In *Paramecium: A current survey*, ed. W. J. vanWagtendonk, 131. Amsterdam: Elsevier.

10. Barnett, A., and Steers, E. Jr. 1984. Antibody-induced membrane fusion in *Paramecium*. *J. Cell Sci.* 65:153.
11. Eisenbach, L., Ramanathan, R., and Nelson, D. L. 1983. Biochemical studies of the excitable membrane of *Paramecium tetraurelia*. IX. Antibodies against ciliary membrane proteins. *J. Cell Biol.* 97:1412.
12. Capdeville, Y., et al. 1987. Allelic antigen and membrane-anchor epitopes of *Paramecium primaurelia* surface antigens. *J. Cell Sci.* 88:553.
13. Beale, G. H. and Kacser, H. 1957. Studies on the antigens of *Paramecium aurelia* with the aid of fluorescent antibodies. *J. Gen. Microbiol.* 17:68.
14. Beale, G. H. and Mott, M. R. 1962. Further studies on the antigens of *Paramecium aurelia* with the aid of fluorescent antibodies. *J. Gen. Microbiol.* 26:617.
15. Mott, M. R. 1963. Cytochemical localization of antigens of *Paramecium* by ferritin-conjugated antibody and by counterstaining the resultant absorbed globulin. *J. R. Microscop. Soc.* 81:159.
16. Mott, M. R. 1965. Electron microscopy studies on the immobilization antigens of *Paramecium aurelia*. *J. Gen. Microbiol.* 41:251.
17. Merkel, S. J., Kaneshiro, E. S., and Gruenstein, E. I. 1981. Characterization of the ciliary membrane proteins of wild-type *Paramecium tetraurelia* and a *pawn* mutant. *J. Cell Biol.* 189:206.
18. Macindoe, H. and Reisner, A. H. 1967. Adsorption titration as a specific semi-quantitative assay for soluble and bound *Paramecium* serotypic antigen. *Aust. J. Biol. Sci.* 20:141.
19. Ramanathan, R., Adoutte, A., and Dutte, R. R. 1981. Biochemical studies of the excitable membrane of *Paramecium tetraurelia*. V. Effects of proteases on the ciliary membrane. *Biochim. Biophys. Acta* 188:4637.
20. Harumoto, T. and Miyake, A. 1992. Possible participation of surface antigens of *Paramecium* in predator-prey interaction. *J. Protozool.* 39:47A.
21. Harumoto, T. 1994. Predator-prey interaction in ciliates: Organelles and molecules participating in the interaction. In *Progress in protozoology*, ed. K. Hausmann and N. Hülsmann, 55. Stuttgart: Fischer-Verlag.
22. Paquette, C. A., et al. 2001. Glycophosphatidylinositol-anchored proteins in *Paramecium tetraurelia*: Possible role in chemoresponse. *J. Exp. Biol.* 204:2899.
23. Yano, J., Rachochy, V., and van Houten, J. L. 2003. Glycophosphatidylinositol-anchored proteins in chemosensory signalling: Antisera manipulation of *Paramecium tetraurelia PIG-A* gene expression. *Eukaryot. Cell* 2:1211.
24. Preer, J. R. Jr. 1959. Studies on the immobilization antigens of *Paramecium*. I. Assay methods. *J. Immunol.* 83:276.
25. Finger, I. 1956. Immobilizing and precipitating antigens of *Paramecium*. *Biol. Bull.* 111:358.
26. Beale, G. H. 1948. The process of transformation of antigenic type in *Paramecium aurelia*, variety 4. *Proc. Natl. Acad. Sci. U.S.A.* 34:418.
27. Sonneborn, T. M. and LeSuer, A. 1948. Antigenic characters of *P. aurelia* (variety 4), determination, inheritance and induced mutations. *Am. Nat.* 82:69.
28. Allen, S. L., et al. 1998. Proposed genetic nomenclature rules for *Tetrahymena thermophila, Paramecium primaurelia* and *Paramecium tetraurelia*. *Genetics* 149:459.
29. Beale, G. H. 1952. Antigen variation in *P. aurelia*, variety 1. *Genetics* 37:62.
30. Margolin, P. 1956. An exception to mutual exclusion of the ciliary antigens in *Paramecium aurelia*. *Genetics* 41:685.
31. Preer, J. R. Jr. 1986. Surface antigens of *Paramecium*. In *The molecular biology of ciliated protozoa*, ed. J. G. Gall, 301. New York: Academic Press.
32. Sonneborn, T. M. 1950. The cytoplasm in heredity. *Heredity* 4:11.
33. Sonneborn, T. M. 1975. The *Paramecium aurelia* complex of fourteen sibling species. *Trans. Am. Microsc. Soc.* 94:155.

34. Sonneborn, T. M. 1951. The role of the genes in cytoplasmic inheritance. In *Genetics in the 20th century*, ed. L. C. Dunn, chap. 14. New York: Macmillan.
35. Reisner, A. 1955. A method for obtaining specific serotype mutants in *Paramecium aurelia*, stock 169, var. 4. *Genetics* 40:591.
36. Preer, J. R. Jr., Preer, L. B., and Rudman, B. 1981. mRNAs for the immobilization antigens of *Paramecium*. *Proc. Natl. Acad. Sci. U.S.A.* 78:6776.
37. Forney, J. D., et al. 1983. Structure and expression of genes for surface proteins in *Paramecium*. *Mol. Cell. Biol.* 3:466.
38. Jones, I. G. and Beale, G. H. 1963. Chemical and immunological comparisons of allelic immobilization antigens in *Paramecium aurelia*. *Nature* 196:205.
39. Sonneborn, T. M. 1945. Gene action in *Paramecium*. *Ann. Mo. Bot. Garden* 32:213.
40. Sonneborn, T. M. 1947. Recent advances in the genetics of *Paramecium* and *Euplotes*. *Adv. Genet.* 1:264.
41. Beale, G. H. 1951. Nuclear and cytoplasmic determinants of hereditary characters in *Paramecium aurelia*. *Nature* 167:256.
42. Beale, G. H. 1954. *The genetics of* Paramecium aurelia. London: Cambridge University Press.
43. Delbrück, M. 1949. Discussion in "Influence des gènes, des plasmagènes et du milieu dans le déterminisme des caractères antigéniques chez *Paramecium aurelia* (variété 4)." In *Unités biologiques douées de continuité génétique*, ed. A. Lwoff, 25, 33. Paris: Colloques Internationaux du CNRS.
44. Monod, J. and Jacob, F. 1962. General conclusions: Teleonomic mechanisms in cellular metabolism, growth, and differentiation. *Cold Spring Harbor Symp. Quant. Biol.* 261:389.
45. Austin, M. L. 1959. The effect of high and low pH and of gelating and liquefying agents on antigenic transformation in *Paramecium aurelia*. *Science* 130:1412.
46. Austin, M. L. 1963. Progress in control of the emergence and the maintenance of serotypes of stock 51, variety 4, of *Paramecium aurelia*. *J. Protozool.* 10(Suppl.):21.
47. Austin, M. L. 1963. The influence of an exchange of genes determining the D antigens between stocks 51 and 32, syngen 4, *Paramecium aurelia*, on the transformation of D to B and A in patulin. *Genetics* 48:881.
48. Austin, M. L., Widmayer, D., and Walker, L. M. 1956. Antigenic transformation as adaptive response of *Paramecium* to patulin; relation to cell division. *Physiol. Zool.* 29:261.
49. Jollos, V. 1921. Experimentelle Protistenstudien. I. Untersuchungen über Variabilität und Vererbung bei Infusorien. *Arch. Protistenk.* 43:1.
50. Sonneborn, T. M. 1950. Heredity, environment and politics. *Science* 111:529.
51. Beale, G. H. 1948. The process of transformation of antigenic type in *Paramecium aurelia*, variety 4. *Proc. Natl. Acad. Sci. U.S.A.* 34:418.
52. Skaar, D. 1956. Past history and pattern of serotype transformation in *Paramecium aurelia*. *Exp. Cell Res.* 10:646.
53. Sonneborn, T. M. 1975. *Paramecium aurelia*. In *Handbook of genetics*, ed R. C. King, chap. 20. Vol. 2. New York: Plenum Press.
54. Pringle, C. R. 1956. Antigenic variation in *Paramecium aurelia*, variety 9. *Z. Indukt. Abstammungs-Vererbungsl.* 87:421.
55. Pringle, C. R. and Beale, G. H. 1960. Antigenic polymorphism in a wild population of *Paramecium aurelia*. *Genet. Res.* 1, 62.
56. Beale, G.H. 1957. The antigen system of *Paramecium aurelia*. *Int. Rev. Cytol.* 6:1.
57. Epstein, L. M., and Forney, J. D. 1984. Mendelian and non-Mendelian mutations affecting surface antigen expression in *Paramecium tetraurelia*. *Mol. Cell. Biol.* 4:1583.
58. Sonneborn, T. M. 1948. The determination of hereditary antigenic differences in genetically identical *Paramecium* cells. *Proc. Natl. Acad. Sci. U.S.A.* 34:413.
59. Beale, G. H. 1952. Gene control of gene expression in *Paramecium aurelia*. *Science* 115:480.

60. Capdeville, Y. 1979. Regulation of surface antigen expression in *Paramecium aurelia*. II. Role of the surface antigen itself. *J. Cell. Physiol.* 99:38.
61. Sonneborn, T. M., Ogasawara, F., and Balbinder, E. 1953. The temperature sequence of the antigenic type in variety 4 of *Paramecium aurelia* in relation to the stability and transformation of antigenic types. *Microb. Genet. Bull.* 7:27.
62. Caron, F. and Meyer, E. 1989. Molecular basis of surface antigen variation in *Paramecia*. *Annu. Rev. Microbiol.* 43:23.
63. Jones, I. G. 1965. Immobilization antigens in heterozygous clones of *Paramecium aurelia*. *Nature* 207:769.
64. Dippell, R. 1953. Serotype expression in heterozygotes of variety 4, *Paramecium aurelia*. *Microbial Genet. Bull.* 7:12.
65. Capdeville, Y. 1971. Allelic modulation in *Paramecium aurelia* heterozygotes in syngen 1. *Mol. Gen. Genet.* 112:306.
66. Capdeville, Y., Vierny, C., and Keller, A.-M. 1978. Regulation of surface antigen expression in *Paramecium primaurelia*. Genetic and physiological factors involved in allelic exclusion. *Mol. Gen. Genet.* 161:23.
67. Preer, J. R. Jr. 1959. Studies on the immobilization antigens of *Paramecium*. II. Isolation. *J. Immunol.* 83:378.
68. Preer, J. R. Jr. 1959. Studies on the immobilization antigens of *Paramecium*. III. Properties. *J. Immunol.* 83:385.
69. Preer, J. R. Jr. 1959. Studies on the immobilization antigens of *Paramecium*. IV. Properties of the different antigens. *Genetics* 44:803.
70. Merkel, S. J., Kaneshiro, E. S., and Gruenstein, E. I. 1981. Characterization of the cilia and ciliary membrane proteins of wild-type *Paramecium tetraurelia* and a *pawn* mutant. *J. Cell. Biol.* 189:206.
71. Capdeville, Y., Cardoso de Almeida, M. J., and Deregnaucourt, C. 1987. The membrane anchor of *Paramecium* temperature-specific surface antigens is a glycosylinositol phospholipid. *Biochem. Biophys. Res. Commun.* 147:1219.
72. Steers, E. Jr. 1962. A comparison of the tryptic peptides obtained from immobilization antigens of *Paramecium aurelia*. *Proc. Natl. Acad. Sci. U.S.A.* 48:867.
73. Steers, E. Jr. 1965. Amino acid composition and quaternary structure of an immobilization antigen from *Paramecium aurelia*. *Biochemistry* 4:1896.
74. Jones, I. G. 1965. Studies on the characterization and structure of the immobilization antigens of *Paramecium aurelia*. *Biochem. J.* 96:17.
75. Reisner, A. H., Rowe, J., and Macindoe, H. 1968. Structural studies on the ribosomes of *Paramecium*: Evidence for a primitive animal ribosome. *J. Mol. Biol.* 32:587.
76. Reisner, A. H., Rowe, J., and Sleigh, R. 1969. The largest known monomeric globular protein. *Biochim. Biophys. Acta* 188:196.
77. Finger, I., et al. 1966. Biosynthesis and structure of *Paramecium* hybrid antigen. *J. Mol. Biol.* 17:86.
78. Hansma, H. G. 1975. The immobilization antigen of *Paramecium aurelia* is a single polypeptide chain. *J. Protozool.* 22:257.
79. Davis, R. H. Jr. and Steers, E. Jr. 1978. Purification of the i-antigen 51A from *Paramecium tetraurelia* by immunoaffinity chromatography. *Immunochemistry* 15:371.
80. Laemmli, U. K. 1970. Cleavage of structural proteins during the assembly of bacteriophage T4. *Nature* 227:680.
81. Meyer, E., Caron, F., and Baroin, A. 1985. Macronuclear structure of the *G* surface antigen gene of *Paramecium primaurelia* and direct expression of epitopes in *Escherichia coli*. *Mol. Cell. Biol.* 5:2414.
82. Prat, A., et al. 1986. Nucleotide sequence of the *Paramecium primaurelia* G surface protein: A huge protein with a highly periodic structure. *J. Mol. Biol.* 189:47.

83. Prat, A. 1990. Conserved sequences flank variable tandem repeats in two alleles of the G surface protein of *Paramecium primaurelia*. *J. Mol. Biol.* 211:521.
84. Bourgain-Guglielmetti, F. and Caron, F. 1996. Molecular characterization of the D surface protein gene subfamily in *Parameciun primaurelia*. *J. Eukaryot. Microbiol.* 43:303.
85. Scott, J., Leeck, C., and Forney, J. 1994. Analysis of the micronuclear surface protein gene in *Paramecium tetraurelia*. *Nucl. Acids Res.* 22:5079.
86. Preer, J. R. Jr. et al. 1987. Molecular biology of the genes for immobilization antigens in *Paramecium*. *J. Protozool.* 34:418.
87. Breuer, M., et al. 1996. Molecular characterization of the D surface protein gene subfamily in *Paramecium tetraurelia*. *J. Eukaryot. Microbiol.* 43:314.
88. Godiska, R. 1987. Structure and sequence of the *H* surface protein gene of *Paramecium* and comparison with related genes. *Mol. Gen. Genet.* 208:529.
89. Caron, F. and Meyer, E. 1985. Does *Paramecium primaurelia* use a different genetic code in its macronucleus? *Nature* 314:185.
90. Preer, J. R. Jr. et al. 1985. Deviation from the universal code shown by the gene for surface protein 51A in *Paramecium*. *Nature* 314:188.
91. Helftenbein, E. 1985. Nucleotide sequence of a macronuclear DNA molecule coding for alpha tubulin from the ciliate *Stylonychia lemnae*. Special codon usage TAA is not a translation termination codon. *Nucl. Acids Res.* 13:415.
92. Horowitz, S. and Gorovsky, M. A. 1985. An unusual genetic code in nuclear genes of *Tetrahymena*. *Proc. Natl. Acad. Sci. U.S.A.* 82:2452.
93. Prescott, D. M. 1994. The DNA of ciliated protozoa. *Microbiol. Rev.* 58:233.
94. Caron, F. 1990. Eukaryotic codes. *Experientia* 46:1106.
95. Tourancheau, A.B., et al. 1995. Genetic code deviations in the ciliates: Evidence for multiple and independent events. *EMBO J.* 14:3262.
96. Liang, A., Schmidt, H. J., and Heckmann, K. 1994. The alpha- and beta-tubulin genes of *Euplotes octocarionatus*. *J. Eukaryot. Microbiol.* 41:163.
97. Nielsen, E., You, Y., and Forney, J. 1991. Cysteine residue periodicity is a conserved structural feature of variable surface proteins from *Paramecium tetraurelia*. *J. Mol. Biol.* 222:835.
98. Caron, F. and Ruiz, F. 1992. A method for the amplification of *Paramecium* micronuclear DNA by polymerase chain reaction and its application to the central repeats of *Paramecium primaurelia* G surface protein. *J. Protozool.* 39:312.
99. Capdeville, Y. 1979. Intergenic and interallelic exclusion in *Paramecium primaurelia*: Immunological comparisons between allelic and non-allelic surface antigens. *Immunogenetics* 9:77.
100. Capdeville, Y. 1969. Sur les interactions entre allèles contrôlant le type antigénique G chez *Paramecium aurelia*. *C. R. Acad. Sci.* (Paris) 269:1213.
101. Schmidt, H. J., et al. 1988. Characterization of *Caedibacter* endonucleobionts from the macronucleus of *Paramecium caudatum* and the identification of a mutant with blocked R-body synthesis. *Exp. Cell Res.* 174:49.
102. Caron, F. and Meyer, E. 1989. Molecular basis of surface antigen variation in *Paramecium*. *Annu. Rev. Microbiol.* 43:23.
103. Martin, L. D., et al. 1994. DNA sequence requirements for the regulation of immobilization antigen A expression in *Paramecium tetraurelia*. *Dev. Genet.* 15:443.
104. Leeck, C. L. and Forney, J. D. 1994. The upstream region is required but not sufficient to control mutually exclusive expression of *Paramecium* surface genes. *J. Biol. Chem.* 269:31283.
105. Leeck, C. L. and Forney, J. D. 1996. The 5′ coding region of *Paramecium* surface antigen genes controls mutually exclusive transcription. *Proc. Natl. Acad. Sci. U.S.A.* 93:2838.

106. Schmidt, H. J. 1987. Characterization and comparison of genomic DNA clones containing complementary sequences to mRNA from serotype 51D of *Paramecium tetraurelia*. *Mol. Gen. Genet.* 209:450.
107. Kimball, R. F. 1964. Physiological genetics of the cilates. In *Biochemistry and physiology of protozoa*, ed. S. H. Hutner, 243. Vol. 3. New York: Academic Press.
108. Borst, P. and Greaves, D. R. 1987. Programmed gene rearrangement altering gene expression. *Science* 235:658.
109. Pays, E. and Steinert, M. 1988. Control of antigen gene expression in African trypanosomes. *Annu. Rev. Genet.* 22:107.
110. Keller, A. M., et al. 1992. The differential expression of the G surface antigen alleles in *Paramecium primaurelia* heterozygous cells correlates to macronuclear DNA rearrangement. *Dev. Genet.* 13:306.
111. Antony, C. and Capdeville, Y. 1989. Uneven distribution of surface antigens during antigenic variation in *Paramecium primaurelia*. *J. Cell Sci.* 92:205.
112. Sinden, R. E. 1973. The synthesis of immobilization antigen in *Paramecium aurelia* in ribosomal cell fractions. *J. Protozool.* 20:307.
113. Sommerville, J. 1968. Immobilization antigen synthesis in *Paramecium aurelia*. Labelling antigen *in vivo*. *Exp. Cell Res.* 56:660.
114. Sommerville, J. 1969. Serotype transformation in *Paramecium aurelia*. *Exp. Cell Res.* 57:443.
115. Meyer, E., Caron, F., and Guiard, B. 1984. Blocking of *in vitro* translation of *Paramecium* messenger RNAs is due to messenger RNA primary structure. *Biochimie* 66:403.
116. Gilley, D., et al. 1990. Multilevel regulation of surface antigen gene expression in *Paramecium tetraurelia*. *Mol. Cell. Biol.* 10:1538.
117. Godiska, R. et al. 1987. Transformation of *Paramecium* by microinjection of a cloned serotype gene. *Proc. Natl. Acad. Sci. U.S.A.* 84:7590.
118. Gilley, D., et al. 1988. Autonomous replication and addition of telomere-like sequences to DNA micro-injected into *Paramecium tetraurelia* macronuclei. *Mol. Cell. Biol.* 8:4765.
119. Jessop-Murray, H., et al. 1991. Permanent rescue of a non-Mendelian mutation of *Paramecium* by microinjection of specific DNA sequences. *Genetics* 129:727.
120. Kim, C. S., Preer, J. R. Jr., and Polisky, B. 1992. Bacteriophage lambda DNA fragments replicate in the *Paramecium* macronucleus: Absence of active copy number control. *Dev. Genet.* 12:97.
121. Bourgain, F. M. and Katinka, M. D. 1991. Telomeres inhibit end to end fusion and enhance maintenace of linear DNA molecules injected into the *Paramecium primaurelia* macronucleus. *Nucl. Acids Res.* 19:1541.
122. Haynes, W. J. et al. 1995. Induction of antibiotic resistance in *Paramecium tetraurelia* by the bacterial gene *H-AP3'-II*. *J. Eukaryot. Microbiol.* 42:83.
123. Thai, K. Y. and Forney, J. D. 2000. Analysis of the conserved cysteine periodicity of *Paramecium* variable surface antigens. *J. Eukaryot. Microbiol.* 47:242.

8 Micronuclei and Macronuclei

8.1 PLOIDY LEVELS

As we have seen, the diploid micronucleus undergoes mitosis at binary fission and meiosis at autogamy and conjugation. Measurement of the relative amounts of DNA in the micronucleus and macronucleus of *P. tetraurelia* has been done on individual cells using a microspectrophotometer after Feulgen staining, which is specific for DNA.[1] These data were confirmed by Woodard et al.[2] and produced a value of 430 times as much DNA in the macronucleus as in the micronucleus, or 860n for the macronucleus and 2n for the micronucleus. Similar results were obtained by Gibson and Martin[3] for various species in the *P. aurelia* complex. Thus, while the micronucleus appears to be diploid, the macronucleus is highly polyploid. Micronuclear polyploidy, as well, has been demonstrated for several species of paramecia. See the review by Raikov.[4]

8.2 ACTIVITY OF MICRONUCLEI AND MACRONUCLEI

It is generally held that the macronuclei contain the active genes responsible for life of paramecia, and that the micronuclei contain inactive genes, which perform a generative function at autogamy and conjugation. This conclusion is supported by a number of kinds of evidence. Morphological studies show that the macronucleus contains numerous nucleoli that are filled with ribosomes, but that the micronuclei lack them. This is taken as evidence that the macronuclear genes are being expressed.

It is also found that amacronucleate cells either die without division or divide no more than once after losing their macronuclei.[5] Amicronucleate cells occur naturally, especially in some mutant lines, such as cells homozygous for the *am* gene.[6] Amicronucleate cells divide normally until autogamy or conjugation, which they are unable to complete. When cells are fed radioactive uridine (a constituent of RNA, but not DNA), the micronuclei and macronuclei become labeled.[7] Although the micronuclei synthesize RNA, this RNA has not been shown to be mRNA, so one cannot know whether the micronuclear genes are producing proteins or not.

The conclusion that the macronucleus contains active genes that control the phenotype of the cell, while the micronucleus has only a generative function, is also supported by the construction of heterocaryons. A heterocaryon is a cell containing different genes in the micronucleus and macronucleus. Remember from Chapter 3 that in macronuclear regeneration at conjugation it is possible to get lines derived from the macronuclear fragments, which are normally lost after conjugation. If a cross is made between clones having a homogyzous dominant gene (*AA*) and a strain homozygous for its recessive allele (*aa*) and at the same time macronuclear regeneration is induced, then all the progeny will be *Aa* in their micronuclei. On the other hand, the progeny derived from macronuclear fragments of the *aa* parent will have *aa* macronuclei, while progeny derived from the *AA* exconjugant will be *AA* in their macronuclei. Thus, the genotypes will be different in the micronuclei (*Aa*) and macronuclei (*AA* and *aa*). Heterocaryons have been produced for mating types,[8,9] kappa maintenance genes,[10] serotype genes,[11] *pwA*, a behavioral mutant,[12] and the *M* gene, for maintenance of the symbiont, mu.[7] The phenotype of such individuals whose macronuclei derive from the homozygous recessive parent is found to be that of the recessive macronucleus, in spite of having the dominant gene in the micronucleus. It has been concluded that the micronuclear genes are not expressed. However, since there are so many fewer genes in the micronucleus relative to the macronucleus, the objection can be made that the evidence from heterocaryons is not decisive, for there might not be enough genes in the micronucleus to counter the big excess of genes in the macronucleus. Nevertheless, Mikami[13] notes that heterocaryons prepared by whole nucleus transplantation have been studied in *P. caudatum* (genes for trichocyst discharge, *tnd-1* and *tnd-2*, and ciliary reversal, *cnrA*, *cnrB*, *cnrC*, *cnrD*). The examination of timed stages during conjugation (normal conjugation, not macronuclear regeneration) between such heterokaryons showed that expression of new macronuclear wild-type genes occurs before the gene level rises above one or two copies, thus making this objection irrelevant.

8.3 PROGRAMMED CHROMOSOMAL BREAKS AND TELOMERES

The fragments of DNA in the macronucleus, like chromosomes, divide at cell division, maintain their integrity, and have telomeres. However, they lack centromeres and do not undergo mitosis like typical chromosomes. Nevertheless, we will call them macronuclear chromosomes. The haploid number of micronuclear chromosomes in stock 51 is estimated by Dippell[14] to be about 40n to 45n. The amount of DNA in a genome uncorrected for GC content is estimated as 90,000 kb[15,16] based on renaturing kinetics. Dividing the amount of DNA in one genome by 40 to 45, one comes up with the average size of a micronuclear chromosome as about 2,100 kb. The average size of a macronuclear chromosome, determined by sucrose gradient centrifugation, was found to be only 300 kb.[17] Fragmentation of micronuclear chromosomes is thought to be the main reason for this size difference, with some elimination of DNA also occurring.

Fragmentation in *Paramecium* operates in a complicated and unique fashion; breakage sites show microheterogeneity of the specific base at which a cut is made and macroheterogenity in selecting preferred regions for cuts from a few hundred base pairs up to several thousand base pairs of each other.[18-20] Special sequences of bases at which cuts are made have not been identified. The result, of course, is enormous heterogeneity in the actual sequences of bases found at the ends of macronuclear chromosomes. One explanation for the heterogeneity is that breaks occur at a distant site and then are cut out upstream from the cut site for a variable number of bases by an exonuclease. However, Coyne et al.[21] point out that this hypothesis does not fit well with the finding of Gilley et al.[22] that new telomeres are readily added onto the terminal base of injected fragments of all foreign DNAs without any additional deletions from the end.

In *Tetrahymena* the chromosome breaks a few bases from the 15 bp chromosome breakage sequence (Cbs), 5'AAAGAGGTTGGTTTA3'.[23] Numerous Cbs are found in the micronuclear genome, and when a new macronucleus is formed, cuts occur at each Cbs, eliminating the Cbs and base pairs 5 to 25 on either side of the Cbs, creating a small amount of heterogeneity.[24] Approximately 15% of the DNA is lost at this time.[25] At macronuclear formation in hypotrichs the DNA is cut into numerous pieces, each containing a single gene, and a very large portion of the micronuclear DNA is discarded (reviewed in Prescott[26]).

Telomeres are found on the ends of the macronuclear chromosomes. In *Paramecium*, as in many other organisms, telomeres may vary in length during cell replication. They consist of repeats of GGGGTT (G_4T_2) in *Tetrahymena*, while in *Paramecium* they consist of about 70% G_4T_2 and 30% G_3T_3. Telomeres serve the function of sealing the ends of the chromosomes so they cannot anneal with each other, as happens in higher organisms in the breakage-fusion bridge cycle.[27,28]

A telomerase in *Tetrahymena* has been isolated and is well characterized.[29-31] This enzyme extends the ends of telomeres, G_4T_2. The enzyme contains a piece of RNA with the complementary sequence of C_4A_2. Telomere length has functional significance in various organisms and has been studied extensively.[31] As noted above, the telomere in *Paramecium* is also a stretch of G_4T_2, but with an occasional G_3T_3 mixed in. One might expect that the polymerase in *Paramecium* consists of two enzymes, one containing C_4A_2 and one containing C_3A_3, but this proves not to be the case. The telomerase in *Paramecium* has been isolated[32,33] and is much like the one in *Tetrahymena*, except it contains C_4A_2, which often makes mistakes and puts in G_3T_3 instead of G_4T_2. Telomeres are added with regularity in *Paramecium*, for as noted above it has been found[34] that the injection of virtually any piece of DNA, even fragments of lambda, is rapidly telomerized and replicated. (It is also noteworthy that the problem of reproducing both strands of DNA at the ends of the chromosome is solved by the action of telomerase.[31])

8.4 INTERNAL ELIMINATED SEQUENCES

Internal eliminated sequences (IESs) are approximately twenty to several thousand base pairs long. They are present in the micronucleus but are eliminated when the micronucleus produces the macronucleus. For a recent review of IESs, see Jahn and Klobutcher.[35] First discovered in *Tetrahymena* by Yao et al.[36] and Callahan et al.,[37] they are classified into two groups: (1) large, transposon-like elements and (2) small elements.

The large, transposon-like elements are 4 to 5 kb in size. They include the TBE1 elements of *Oxytricha*,[38,39] the Tec1 and Tec2 elements of *Euplotes*,[40-42] and the Tel-1 and Tlr elements of *Tetrahymena*.[43,44] They have a dozen or so open reading frames that include a transposase. This enzyme is similar in sequence to the transposase of certain known transposons.[45] At conjugation and autogamy the large transposon-like IESs are removed from the macronucleus.

The small IESs vary in size from a few bases up to a kilobase. They do not contain open reading frames. At conjugation they are removed from the genome.

Many thousands of these small IESs are present in the genome. As noted above, they were first discovered in *Tetrahymena* in Gorovsky's laboratory,[36] and Yao has studied them extensively. They are also found in hypotrichs[46] and in *Paramecium*.[47] In *Paramecium* and in hypotrichs there is a TA on both ends in the micronuclear copy of the IES. When the macronucleus is formed, breaks occur between the two TAs, and after elimination of the IES, a single TA is left in the copy of DNA destined for the macronucleus.

Thus, IES elimination in *Paramecium* is exact, and IESs are often found in the reading frame of genes. When they are eliminated, the reading frame is always precisely maintained.[47] In *Tetrahymena* IESs are not precise and they are not terminated with TA. After their discovery in the immobilization antigens of *Paramecium* they were found to be quite numerous: seven in the coding region of the 51A gene and several IESs outside the transcribed portion of the gene. In the 51B gene they often occur in the same place, but differences are also found. The sequences internal to the TAs are similar, with inverted repeat sequences at either end of the IESs. A consensus sequence has been described for *Paramecium* and *Euplotes*, 5'-TAYAG-YNR-3' (Y = pyrimidine, R = purine, N = any base), that appears on each end of the IESs.[48] The ends of the Tec transposons of *Euplotes* are similar; all are eliminated when the micronuclei form macronuclei.

Mention must also be made of the IESs in hypotrichs studied by Prescott.[49] In these cases, as always, the IESs separate the macronuclear destined sequences, but it turns out that the segments of the macronuclear destined sequences are scrambled and some even inverted. Of course, they must be unscrambled and oriented properly before a functional macronuclear gene can be formed. Precisely how that is accomplished is still not known, although models have been proposed.

Mutations within IESs were found by Forney and students. See Matsuda et al.[50] for a review of gene mutations that block IES removal. In these mutants the IES is not removed at autogamy and the resulting clones do not express the gene in which they are found due to frameshifts introduced by inclusion of the IES. A total of six genes were listed, occurring at different points in the consensus sequence, 5'-TAYAGYNR-3'; they included not only mutations in the serotype A genes, *aim 1* (A gene IES mutation), *aim3*, *aim4*, *aim5*,[50–52] but also a mutant in pawn B (*pwB662*)[53] and in a tubulin gene (*sm19-1*)[54] Four of these mutations occurred in either position 1 or 2 (T or A) in the consensus sequence, and the other two at position 6. The mutations all blocked IES elimination. One of these was interesting in that it showed that one IES actually had another IES within it.[51] It has also been shown that bases immediately flanking, or up to fifty bases away from an IES, are necessary for proper removal of IESs.[55]

The suggestion has been made that IESs might have been evolved from the transposon-like forms.[56] One assumes that the essential functions of the transposons were transferred to the genome of the host ciliate, and many of the properties of IESs support this view.

Different models of IES excision have been proposed for the different IESs in different organisms. In *Paramecium* it is thought that during IES elimination double breaks occur, and then circles are produced secondarily before being lost. The mechanism of IES excision has been studied in *Paramecium*,[57] and the reader is referred to Gratias and Betermier[57] for data and models.

8.5 INDUCED MENDELIAN DELETIONS

A series of x-ray-induced Mendelian deletions has been found in the *A* gene in stock 51 of *P. tetraurelia* beginning at different sites, approximately −1300 (d12), −1000 (d1), +1 (d1, d12), +8300 (d16), and extending to the end of the chromosome.[58,59] These mutants all produce typical genic 1:1 ratios in autogamous F2 lines derived from crosses to wild type.[59] It was also inferred on the basis of polymerase chain reaction (PCR) and the observation that the larger deletion mutants had given rise to smaller deletion mutants, that the whole gene was present in the micronucleus of each one of the mutants. However, it was later decided that this conclusion was incorrect for when micronuclei became available, direct blots of d12 micronuclear DNA showed not the whole gene, but only the fragment of the gene above the cut site.[60] Presumably, the PCR result was produced by contamination. However, the evidence indicating that d12 (−1300) and d1 (−1000) gave rise to less extensive deletions d12 (+1) and d1 (+1) is more difficult to interpret, but it may be due to the fact that different deletions produced mixed cultures in the first isolations, and that these mixed lines were later sorted out from mass cultures. In any case, the alleles with break points at −1300, −1000, and +1 are now stable deletions and behave in a Mendelian fashion as separate alleles at the *A* locus when crossed to wild type or to each other.

Two unlinked loci, designated d8 and d29, unlinked to the serotype *A* gene, have been shown to affect the stability of expression of the *A* gene in *P. tetraurelia*.[58] Thus, genes other than the *A* locus are important in determining which serotype is expressed.

8.6 MAINTAINING GENIC BALANCE AT MACRONUCLEAR DIVISION

Genetic balance is maintained in multicellular organisms by mitosis, in which each chromosome divides once in each interfission interval and the daughters are directed by spindle fibers to one of the new cells. Since this process produces an equal number of all the genes in a genome, yet proteins are needed in different amounts, promoter efficiency and activation and repression of the activity of genes determine that proteins are produced in amounts that meet the needs of the cell. However, in the case of macronuclear chromosomes, no evidence of centromeres or mitosis has ever been found. It was once thought that the chromosomes were held together in haploid or diploid subunits,[9] and that at cell division the subunits divided and were distributed to the daughter cells mitotically, thus solving the problem of maintaining genic balance. This hypothesis was considered likely for many years. Although some microscopical evidence was thought to support this hypothesis, much of the microscopical evidence did not.[61] The findings that macronuclear chromosomes lack a centromere and that each chromosome ends in a telomere (see next section) have more recently made the hypothesis of subunits that require hidden mitoses less attractive. Moreover, the conceptual problem of how to unite the extremely numerous gene-sized macronuclear chromosomes in the hypotrichs into subunits constitutes still another difficulty with the subunit hypothesis.

If subunits do not exist, and the macronuclear chromosomes are distributed in a random fashion at cell division, then some aneuploid imbalances must occur.

Kimball[62] might have been the first to produce evidence that the total amount of macronuclear DNA in individual paramecia continuously varies, while populations of cells always tend to have the same average amount. This fact was recognized clearly by Berger and Schmidt,[63] who showed that the two daughter macronuclei formed at fission definitely have different amounts of total DNA. They produced evidence that if a cell had too little DNA, then the total DNA was slowly regulated upward, and if a cell had too much DNA, the total was slowly regulated downward, to maintain a constant average amount. Some molecules must replicate more than once, and others not at all in a cell cycle. In that event, some mechanism of control of the replication of individual molecules must exist if genic balance is to be maintained.

Evidence for control is found in *Tetrahymena*, which can live indefinitely without a micronucleus. Perhaps there aneuploid cells arise but are simply eliminated by death. However, the death rate in lines of *Tetrahymena* was shown to be too low to eliminate the large number of aneuploid imbalances produced.[64] It was thought that cells in *P. tetraurelia* senesce and die before they multiply enough to see whether severe imbalance occurs.[65] Senescence that occurs at late stages in the life cycle is thought to have a different basis from that of aneuploidy. But in *Tetrahymena*, some mechanism to maintain balance probably exists. Perhaps the individual macronuclear chromosomes have a mechanism to sense their copy number and increase the likelihood of their dividing twice during a fission interval or not dividing at all, in order to maintain balance.[64,66] A possible mechanism is provided by a consideration of replication control in plasmids.

Studies on plasmid replication have been of great importance in molecular biology, leading to the discovery of antisense RNAs and providing information on the replication and control of gene expression. It is interesting that mixed populations of plasmids maintain numbers characteristic of the individual plasmid in its host bacterium. See review in del Solar et al.[67] It is clear that maintenance of constant numbers of plasmids in the mixed populations of plasmids in bacteria is much like the problem of maintaining constant numbers of mixed chromosomes in the macronucleus of ciliates. There have been numerous studies on this problem in bacteria, and many complicated models have been produced to explain the results of these studies. Balance is maintained because these models all produce more rapid multiplication of plasmids when the concentration of plasmids is low than when the concentration of plasmids is high. In some models the plasmid makes a special protein that reacts with the plasmid and turns its replication off when the number of plasmids and the special proteins reach higher levels. Usually, however, most more recent models suppose that the controlling element is simply a double-stranded copy of plasmid DNA, part of an mRNA (messenger RNA), which is cut into small pieces and acts as RNAi (RNA inhibitor). Since RNAi inhibits protein synthesis by combining with mRNA, it also has been shown to inhibit DNA synthesis when its concentration is high. It does this by hybridization of its antisense strand to the sense strand of DNA near the origin of replication, often with a host of other protein factors. When it is bound in this way, it is found to inhibit replication of plasmids.

In both *Paramecium* and *Tetrahymena* ribosomal DNAs undergo additional amplification, and in hypotrichs some genes are amplified to high levels.[68] Perhaps in the hypotrichs, where usually only a single gene is present on a very small macronuclear chromosome, one might hope to find evidence of individual copy number control, but the problem has not been solved for ciliates as it has been for plasmids.[69]

Since RNAi plays a prominent role in *Paramecium* (see also Chapter 12), it would seem that it might also be the mechanism that controls the number of macronuclear chromosomes, just as it does for plasmids in *E. coli*. All chromosomes would tend to overreplicate when low in concentration, but would be shut down by increasing concentrations of RNAi molecules when the concentration of a particular chromosome reached higher levels. This control would be specific for each particular chromosome and could serve to maintain genic balance at all times. No specific repressor sequences would be required, for the mRNA produced by each gene would contain the sequences of the repressor. If this mechanism operates, then the concentration of a given gene is probably characteristic of that particular gene and might easily produce different levels of specific chromosomes in the macronucleus, such as the very high number of ribosomal genes or the high number of certain genes seen in hypotrichs.

8.7 THE *PARAMECIUM* GENOME PROJECT

In August 1999 at the initiative of the workers at Gif-sur-Yvette, France, a group was formed to sequence the genome of *P. tetraurelia*. In 2002, funds from the CNRS provided financial support, and in 2003, Genoscope approved the total sequencing of the genotype of *P. tetraurelia* strain d4-2 (a derivative of a cross between stocks 29 and 51, containing the allele A^{29} made isogenic with stock 51 by a series of backcrosses to stock 51). The project to sequence the macronuclear genome is now essentially finished.[70]

The results of this work at present include the identification of some 39,642 genes in the entire macronuclear genome of the stock of *Paramecium* analyzed, with very small intergenic regions. These genes appeared in 188 scaffolds (equivalent to macronuclear chromosomes), most found to end in telomeres. The identification of genes was based on the analysis of several mRNA libraries taken at different times during conjugation and autogamy, and comparisons with other species to aid in identification of genes. This is a very large genome (39,642 genes), much larger than that of *Tetrahymena* (estimated as 27,424 genes),[71] and the human genome (thought to be about like *Tetrahymena*). The large number of genes in gene families and pseudogenes probably accounts for the large size of the genome.

Evolutionary studies suggest that several times over the evolutionary history of *P. tetraurelia* the genome has undergone duplication of the whole genome. At each time the number of genes rises to double its original value, but then rapidly loses most (but not all) of the duplicate genes, helping to account for the large number of genes.

8.8 NUCLEAR DUALISM

The nuclear dualism of the ciliates is one of their unique features and needs an explanation. Perhaps the system of nuclear dualism might be in response to the level of ploidy required by the large size of the ciliates. The highest chromosome number seen in eukaryotes seems to be in the fern, *Ophioglossum reticulatum*,[72] which has a ploidy level of 1,260. The ploidy level of macronuclei often rises much higher than this. In *P. tetraurelia* the ploidy level is about 1,000, while in the large ciliate, *Spirostomum*, it is estimated at 13,150.[73] See Raikov[4] for a review. The hypothesis that mitosis is difficult or impossible at such high numbers of chromosomes is perhaps reasonable, and it also seems likely that it might be necessary to have many copies of the genome in a very large cell.

REFERENCES

1. Woodard, J., Gelber, B., and Swift, H. 1961. Nucleoprotein changes during the mitotic cycle in *Paramecium aurelia*. *Exp. Cell Res.* 23:258.
2. Woodard, J., et al. 1966. Cytochemical studies of conjugation in *Paramecium aurelia*. *Exp. Cell Res.* 41:55.
3. Gibson, I. and Martin, N. 1971. DNA amounts in the nuclei of *Paramecium aurelia* and *Tetrahymena pyriformis*. *Chromosoma* 35:374.
4. Raikov, I. B. 1982. *The protozoan nucleus*. Vol. 9. Cell Biology Monographs. New York: Springer-Verlag.
5. Wichterman, R. 1985. *The biology of* Paramecium. New York: Plenum Press.
6. Sonneborn, T. M. 1975. *Paramecium aurelia*. In *Handbook of genetics*, ed. R. C. King, chap. 20. Vol. 2. New York: Plenum Press.
7. Pasternak, J. 1967. Differential genic activity in *Paramecium aurelia*. *J. Exp. Zool.* 165:395.
8. Sonneborn, T. M. 1940. The relation of macronuclear regeneration in *Paramecium aurelia* to macronuclear structure, amitosis and genetic determinants. *Anat. Rec. Suppl.* 78:53.
9. Sonneborn, T. M. 1947. Recent advances in the genetics of *Paramecium* and *Euplotes*. *Adv. Genet.* 1, 263.
10. Sonneborn, T. M. 1946. Inert nuclei: Inactivity of micronuclear genes in variety 4 of *Paramecium aurelia*. *Genetics* 31:231.
11. Sonneborn, T. M. 1954. The relation of autogamy to senescence and rejuvenation in *Paramecium aurelia*. *J. Protozool.* 1:38.
12. Berger, J. 1974. Selective autolysis of nuclei as a source of DNA precursors in *Paramecium aurelia* exconjugants. *J. Protozool.* 21:145.
13. Mikami, K. 1988. In *Nuclear dimorphism and function in* Paramecium, ed. H.-D. Görtz, 120. Berlin: Springer-Verlag.
14. Dippell, R. V. 1954. A preliminary report on the chromosomal constitution of certain variety 4 races of *Paramecium aurelia*. *Chromosoma* 53:191.
15. Soldo, A. T. and Godoy, G. A. 1972. The kinetic complexity of *Paramecium* macronuclear deoxyribonucleic acid. *J. Protozool.* 19:673.
16. Cummings, D. J. 1975. Studies on macronuclear DNA from *Paramecium aurelia*. *Chromosoma* 53:191.
17. Preer, J. R. Jr. and Preer, L. B. 1979. The size of macronuclear DNA and its relationship to models for maintaining genic balance. *J. Protozool.* 26:14.
18. Baroin, A., Prat, A., and Caron, F. 1987. Telomeric site position heterogeneity in macronuclear DNA of *Paramecium primaurelia*. *Nucl. Acids Res.* 15:1717.
19. Forney, J. D. and Blackburn, E. H. 1988. Developmentally controlled telomere addition in wild type and mutant paramecia. *Mol. Cell. Biol.* 8:251.
20. Caron, F. 1992. A high degree of macronuclear chromosome polymorphism is generated by variable DNA's rearrangements in *Paramecium primaurelia* during macronuclear differentiation. *J. Mol. Biol.* 225:661.

21. Coyne, R. S., Chalker, D. L., and Yao, M.-C. 1996. Genome downsizing during ciliate development: Nuclear division of labor through chromosome restructuring. *Annu. Rev. Genet.* 30:557.
22. Gilley, D., et al. 1988. Autonomous replication and addition of telomerelike sequences to DNA microinjected into *Paramecium tetraurelia* macronuclei. *Mol. Cell. Biol.* 8:4765.
23. Yao, M.-C., Zheng, K., and Yao, C. H. 1987. A conserved nucleotide sequence at the site of developmentally regulated chromosomal breakage in *Tetrahymena*. *Cell* 48:779.
24. Fan, Q. and Yao, M.-C. 2000. A long stringent sequence signal for programmed chromosome breakage in *Tetrahymena thermophila*. *Nucl. Acids Res.* 28:895.
25. Yao, M.-C. and Gorovsky, M. A. 1974. Comparison of the sequences of the macro- and micronuclear DNA of *Tetrahymena pyriformis*. *Chromosoma* 48:1.
26. Prescott, D. M. 1994. The DNA of ciliated protozoa. *Microbiol. Rev.* 58:233.
27. McClintock, B. 1938. The fusion of broken ends of sister half chromatids following chromatid breakage at meiotic anaphase. *Res. Bull. Mo. Agric. Exp. Stn.* 290:1.
28. Muller, H. J. 1938. The remaking of chromosomes. *Collecting Net* 13:182.
29. Greider, C. W. and Blackburn, E. H. 1987. The telomere terminal transferase of *Tetrahymena* is a ribonuclear protein enzyme with two distinct primer specificity components. *Cell* 51:887.
30. Greider, C. W. and Blackburn, E. H. 1989. A telomeric sequence in the RNA of *Tetrahymena* telomerase required for telomere repeat synthesis. *Nature* 337:331.
31. Chan, R. W. L. and Blackburn, E. H. 2004. Telomeres and telomerase. *Phil. Trans. R. Soc. Lond.* 359B:109.
32. McCormick-Graham, M. and Romero, D. 1996. A single telomerase RNA is sufficient for the synthesis of variable telomeric DNA repeats in ciliates of the genus *Paramecium*. *Mol. Cell. Biol.* 16:1871.
33. Ye, A. J. and Romero, D. P. 2002. A unique pause pattern during telomere addition by the error-prone telomerase from the ciliate *Paramecium tetraurelia*. *Gene* 294:205.
34. Gilley, D. et al. 1988. Autonomous replication and addition of telomerelike sequences to DNA microinjected into *Paramecium tetraurelia* macronuclei. *Mol. Cell. Biol.* 8:4765.
35. Jahn, C. L. and Klobutcher, L. A. 2002. Genome remodeling in ciliated Protozoa. *Annu. Rev. Microbiol.* 56:489.
36. Yao, M.-C., et al. 1984. DNA elimination in *Tetrahymena*: A developmental process involving extensive breakage and rejoining of DNA at defined sites. *Cell* 36:433.
37. Callahan, R. C., Shalke, G., and Gorovsky, M. A. 1984. Developmental rearrangements associated with a single type of expressed alpha-tubulin gene in *Tetrahymena*. *Cell* 36:441.
38. Herrick, G., et al. 1985. Mobile elements bounded by C_4A_4 telomeric repeats in *Oxtricha fallax*. *Cell* 3:459.
39. Williams, K., Doak, T. G., and Herrick, G. 1993. Developmental precise excision of *Oxtricha trifallax* telomere-bearing elements and formation of circles closed by a copy of the flanking target duplication. *EMBO J.* 12:4593.
40. Baird, S. E., et al. 1989. Micronuclear genome organization in *Euplotes crassus*: A transposon-like element is removed during macronuclear development. *Mol. Cell. Biol.* 9:3793.
41. Krikau, M. F. and Jahn, C. L. 1991. Tec2, a second transposon-like element demonstrating developmentally programmed excision in *Euplotes crassus*. *Mol. Cell. Biol.* 11:4751.
42. Jahn, C. L., et al. 1993. Structures of the *Euplotes crassus* Tec1 and Tec2 elements: Identification of putative transposase coding regions. *Gene* 133:71.
43. Wells, J. M., et al. 1994. A small family of elements with long inverted repeats is located near sites of developmentally regulated DNA rearrangement in *Tetrahymena thermophila*. *Mol. Cell. Biol.* 14:5939.

44. Gershan, J. A. and Karrer, K. M. 2000. A family of developmentally excised DNA elements in *Tetrahymena* is under selective pressure to maintain an open reading frame encoding an integrase-like protein. *Nucl. Acids Res.* 28:4105.
45. Doak, T. G., et al. 1994. A proposed superfamily of transposase genes: Transposon-like elements in ciliated protozoa and a common "D35E" motif. *Proc. Natl. Acad. Sci. U.S.A.* 91:942.
46. Ribas-Aparicio, R. M., et al. 1987. Nucleic acid splicing events occur frequently during macronuclear development in the protozoan *Oxtricha nova* and involve the elimination of unique DNA. *Genes Dev.* 1:323.
47. Steele, C. J., et al. 1994. Developmentally excised sequences in micronuclear DNA of *Paramecium*. *Proc. Natl. Acad. Sci. U.S.A.* 91:2255.
48. Klobutcher, L. A. and Herrick, G. 1995. Consensus inverted terminal repeat sequence of *Paramecium* IESs: Resemblance to termini of Tc1-related and *Euplotes* Tec transposons. *Nucl. Acids Res.* 23:2006.
49. Prescott, D. M. 2000. Genome gymnastics: Unique modes of DNA evolution and processing in ciliates. *Nat. Rev. Genet.* 1:191.
50. Matsuda, A., Mayer, K. M., and Forney, J. D. 2004. Identification of single nucleotide mutations that prevent developmentally programmed DNA elimination in *Paramecium tetraurelia*. *J. Eukaryot. Microbiol.* 51:664.
51. Mayer, K. M., Mikami, K., and Forney, J. D. 1998. A mutation in *Paramecium tetraurelia* reveals functional and structural features of developmentally excised DNA elements. *Genetics* 148:139.
52. Mayer, K. M. and Forney, J. D. 1999. A mutation in the flanking 5'-TA-3' dinucleotide prevents excision of an internal eliminated sequence from the *Paramecium tetraurelia* genome. *Genetics* 151:597.
53. Haynes, W. J., et al. 2000. The cloning and molecular analysis of *pawn-B* in *Paramecium tetraurelia*. *Genetics* 155:1105.
54. Ruiz, F., et al. 2000. The *SM19* gene, required for duplication of basal bodies in *Paramecium*, encodes a novel tubulin, eta-tubulin. *Curr. Biol.* 10:1451.
55. Ku, M., Mayer, K,. and Forney, J. D. 2000. Developmentally regulated excision of a 28 base-pair sequence from the *Paramecium* genome requires flanking DNA. *Mol. Cell. Biol.* 20:8390.
56. Klobutcher, L. A. and Herrick, G. 1997. Developmental genome reorganization in ciliated protozoa: The transposon link. *Prog. Nucl. Acid Res. Mol. Biol.* 56:1.
57. Gratias, A. and Betermier, M. 2003. Processing of double-stranded breaks is involved in the precise excision of *Paramecium* internal eliminated sequences. *Mol. Cell. Biol.* 23:7152.
58. Epstein, L. M. and Forney, J. D. 1984. Mendelian and non-Mendelian mutations affecting surface antigen expression in *Paramecium tetraurelia*. *Mol. Cell. Biol.* 4:1583.
59. Rudman, B., et al. 1991. Mutants affecting processing of DNA in macronuclear development in *Paramecium*. *Genetics* 129:47.
60. Preer, J. R. Jr. 2000. Epigenetic mechanisms affecting macronuclear development in *Paramecium* and *Tetrahymena*. *J. Eukaryot. Microbiol.* 47:515.
61. Kimball, R. F. 1953. The structure of the macronucleus of *Paramecium aurelia*. *Proc. Natl. Acad. Sci. U.S.A.* 39:345.
62. Kimball, R. F. 1967. Persistent intraclonal variation in cell dry mass and DNA content in *Paramecium aurelia*. *Exp. Cell Res.* 48:378.
63. Berger, J. D. and Schmidt, H. J. 1978. Regulation of macronuclear DNA content in *Paramecium tetraurelia*. *J. Cell Biol.* 76:116.
64. Preer, J. R. Jr. and Preer, L. B. 1979. The size of macronuclear DNA and its relationship to models for maintaining genic balance. *J. Protozool.* 26:14.
65. Preer, J. R. Jr. 1976. Quantitative predictions of random segregation models of the ciliate macronucleus. *Genet. Res.* (Camb.) 27:227.

66. Brunk, C. F. 1986. Genome reorganization in *Tetrahymena*. *Int. Rev. Cytol.* 99:49.
67. del Solar, G., et al. 1998. Replication and control of circular bacterial plasmids. *Microbiol. Mol. Biol. Rev.* 62:434.
68. Wegner, M., et al. 1989. Identification of an amplification promoting DNA sequence from the hypotrichous ciliate *Stylonychia lemnae*. *Nucl. Acids Res.* 17:8783.
69. Skovorodkin, I. N., et al. 2001. Minichromosomal DNA replication in the macronucleus of the hypotrichous ciliate *Stylonychia lemnae* is independent of chromosome-internal sequences. *Chromosoma* 110:352.
70. Aury, J.-M., et al. 2006. Global trends of whole-genome duplications revealed by the ciliate *Paramecium tetraurelia*. *Nature* 444:171.
71. Eisen, J. A., et al. 2006. Macronuclear genome sequence of the ciliate *Tetrahymena thermophila*, a model eukaryote. *PloS Biol.* e286.
72. Stebbins, G. L. 1966. Chromosomal variation and evolution. *Science* 152:1463.
73. Ovchinnikova, L. P., Selivanova, G. V., and Cheissin, E. M. 1965. Photometric study of the DNA content in the nuclei of *Spirostomum ambiguum* (Ciliata, Heterotricha). *Acta Protozool.* 3:69.

9 Ribosomal RNA and DNA

9.1 THE EARLY WORK OF FINDLY AND GALL

Because ribosomal RNA (rRNA) is present in large amounts in cells it is easily isolated. Because of its ease of isolation and because it plays a key role in protein synthesis, it and the genes that produce it (rDNA) have often been studied. Since it has a rather limited but essential function, it has remained relatively stable over evolutionary time. rRNA and rDNA have a short history in *Paramecium*, having been investigated in only a few laboratories from the late 1970s until the late 1990s. Now it is beginning to emerge again in its usefulness for studies on evolutionary biology. In *Tetrahymena*, however, it has been studied extensively by Meng-Chao Yao and his group, and they have made important additions to our knowledge of the molecular biology of that organism.

The first work on rRNA and rDNA in *P. tetraurelia* stock 51 was carried out by Findly and Gall.[1,2] They isolated rRNA. They also were able to isolate rDNA from the macronucleus readily because they found that it produced a satellite DNA when whole-cell DNA was spun in the presence of actinomycin D and $CsCl_2$. It was present in high concentration. Using restriction enzymes, they mapped the genes and found them to consist of both linear and circular molecules. All the DNA was polymeric, consisting of subunits. Each subunit of the polymer is made up in the order: NTS (nontranscribed spacer), 16s rDNA (coding for small subunit RNA), transcribed spacer 1 (TS1), 5.8s rDNA, transcribed spacer 2 (TS2), and 26s rDNA (large subunit rDNA). See Figure 9.1 for a depiction of the molecule. They found that the number of subunits per polymeric molecule varied, but was generally around a mean of about six. We later found by sequencing (see below) that the linear polymers ended abruptly near the left end of the molecule and with base 500 to 3388 of the NTS, and at the right end with base 1 to 500 of the NTS. Both ends terminated in a telomere of variable length. They used a clever scheme to prove the polymeric nature of the molecules. They partially denatured the DNA to melt it, and then spread it for electron microscopy, before renaturation could occur. Observations showed a pattern of short bubble-like single-stranded regions that had been denatured between long double-stranded sections that had not been melted. The straight regions represented the transcribed DNA, and unpaired regions or bubbles represented the NTS regions. To measure the number of subunits per polymer, one simply counted the number of bubbles per molecule.

Shortly after the work by Findly and Gall, a pulsed-field analysis known as orthogonal-field-alternation gel electrophoresis (OFAGE) was used by Phan et al.[3] to study the organization of rRNA in stock 51. They concluded that while the 51A immobilization antigen gene behaved normally, the rDNA genes behaved in an unusual fashion when the voltage was varied. They concluded that the rDNA genes were circular. Later, however, it was found by Preer et al.[4] that another system for

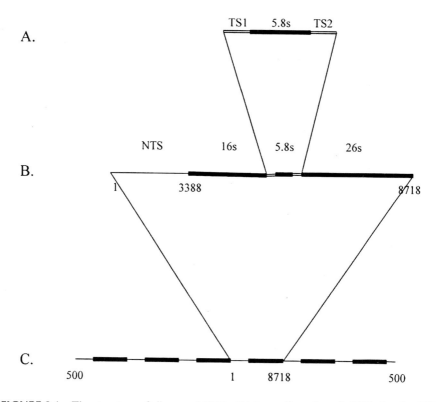

FIGURE 9.1 The structure of ribosomal DNA. (A) A small section of rDNA showing TS1 (transcribed spacer 1), 5.8s rDNA, and TS2 (transcribed spacer 2). (B) A subunit of rDNA or the DNA that codes for one complete unit of RIBX, consisting of bases 1 to 8178. (C) Note that there are several subunits arranged in tandem to produce a molecule of rDNA. These bases end at both ends at about base 500 in the NTS (nontranscribed spacer). Though not pictured here, each end of the polymer is attached to a telomere of variable length.

studying large molecules, the clamped homogeneous electrophoresis field (CHEF) system, gave no differences in migration with different voltages, and the molecules were judged to be linear. It is not clear how the question of the presence or absence of circles is to be resolved.

9.2 MORE RECENT STUDIES

Preer et al.[4] confirmed all the major findings of Findly and Gall. In addition, they isolated 87 different clones, all from stock 51 of *P. tetraurelia*, containing units of macronuclear rDNA, and these clones fell into six different kinds when mapped: the first was called RIBX (48 clones), the second was RIBD (22), the third RIBA (5), the fourth RIBB (4), the fifth RIBC (2), and the sixth RIBE (6). These six kinds of clones (with different subunits) appeared to be identical in the sequences of their transcribed regions, but differed in their nontranscribed spacer regions. The small subunit ribosomal RNA had been sequenced by Sogin and Elwood.[5] Sequences were obtained for the remainder of the molecule, including the spacers of RIBX (3387 bp) and RIBD (2996 bp). It was found that the spacers in these two forms differed by a deletion in a repeated region of 388 bases and 29 base substitutions and single base deletions.

Preer et al.,[4] using stock 51, now turned their attention to the structure of the linear polymers. By using an enzyme that cut just once in a subunit, and sequencing the whole subunit, they found it to have a length of about the size predicted by Findly and Gall. On the assumption that the subunits were tandemly repeated in a linear structure, an attempt was made to locate the end sections. This was done by using enzymes that cut only once in a subunit, but near and at different distances from the putative ends of the molecules. Such cuts liberated three sizes of fragments. First were the internal fragments. Irrespective of the position of the cut, they always produced on sizing gels superimposed fragments of the length of a subunit. The terminal fragments, however, were of two sizes. Moreover, the size of the fragment depended on the position of the enzyme site within the NTS. Since each end was present in only one subunit per molecule, the band was much less prominent. This weak appearance of the terminal section was made even more difficult to see on a sizing gel because of the variability in their position (a few hundred bases) and the fact that they also contained telomeres of variable length. It was concluded that the common break point was within the NTS at approximately base 500 at both the left and right ends of the nontranscribed spacer, section 500 to 8718 base pairs on the left end and 1 to 500 base pairs on the right end. A gel of such a digest showed a large prominent band of subunit length in which the elements are superimposed on each other, and then two other rather fuzzy weaker bands representing the end sections of the whole polymeric molecule. The position of these bands varied, depending on how close the position of the restriction enzyme cut is to the end of the polymer, yet fully confirmed this interpretation.

Final proof of the linear model was found by using polymerase chain reaction (PCR) with one degenerate primer in the telomere (degenerate in order to take care of G4A2 and G3A3 telomeric groups) and the second primer in either end of the macronuclear chromosome. Using these primers it was possible to sequence the telomere and the site where the telomere joined the NTS. The left end of the molecule was found to start in a telomere followed by bases beginning at approximately position 500 and extending to 3388 of the NTS, plus the remainder of the polymer. The right end contained bases beginning at number one of the NTS and extended to approximately 500 and then a telomere at the end. These positions were not precise but varied over a range of 500 plus or minus 200 bases for the beginning and end of the NTS. These numbers were deemed to be consistent with the results obtained from mapping, and also to be important in understanding how these molecules are synthesized.

Dot blots showed that the number of rDNA subunits was amplified in the macronucleus to about twenty thousand, rather than the one thousand for total DNA.

Efforts were also made to examine the micronuclear rDNA genes. Although these genes are not transcribed in the micronucleus, they give rise to the transcribed macronuclear genes at autogamy and conjugation. The micronucleus appears to have only a single copy of each gene. Nineteen clones were isolated from a micronuclear library of stock 51, but they proved to be parts of only four micronuclear genes, RIB1, RIB2, RIB3, and RIB4. Each one coded for a different single macronuclear subunit, but only two, RIB1 and RIB2, were complete. Of these two, RIB1 had a very short spacer that matched no macronuclear gene, but one, RIB2, was complete and mapped like the macronuclear gene RIBX. Presumably several additional micronuclear genes remain to be found and mapped, at least one for each of the macronuclear arrangements RIBX (found), RIBC, and so forth.

9.3 MICRONUCLEAR DNA AND A MODEL

The complete micronuclear clone, RIB2, that mapped like the macronuclear RIBX was sequenced. It was found to have spacer sequences at both ends of the molecule, with most of the spacer on its left end, but the extreme left end was redundant with the extreme right end within the spacer. A model was proposed[4] showing how the left and right ends of the gene could pair in a circular configuration, and then how a rolling circle model of replication could explain the development of the single micronuclear gene into the tandemly repeated linear macronuclear structure observed (Figure 9.2). One also notes that a rolling circle could easily produce circular molecules as well as linear. Since the prediction that the model requires a termination at the same point on both left and right ends, and this is what was found, the model is considered to be rather robust, although the situation with individual molecules is not known. Nevertheless, the individual molecules examined by Findly and Gall show bubbles of varying sizes in individual molecules that would not be predicted by the model. Possibly, bubbles of different sizes might be accounted for by crossing over between molecules of different kinds subsequent to their formation within the macronucleus, or simply by variation in the extent of the denatured regions.

9.4 VARIANTS

The rDNA of seven stocks of *P. tetraurelia*, stocks 47, 51, 127, 148, 172, 203, and 316, were examined by Findly and Gall.[2] They concluded that on the basis of their maps they were all identical except for stock 127, which seemed to differ from the others in its nontranscribed spacer region. Stocks 51 and 127 were examined by Preer et al.[4] They found that the differences observed by Findly and Gall were due to the fact that stock 127 failed to show RIBD. Crosses between stocks 51 and 127 produced a 1:1 segregation of the ability or inability to show RIBD in the F2 by autogamy. They concluded that the gene for RIBD, present in all the stocks, was deleted in stock 127. It should be noted that other differences in the minor types that differ in their restriction maps in the nontranscribed spacer regions might also exist, but would not have been detected by blots of whole-cell DNA.

As noted above, the rDNA of the *Paramecium aurelia* complex appears to be relatively stable throughout evolution in its small and most of its large subunits, and these subunits have been used extensively in trying to assess the evolutionary relationships between members of the complex and other morphological species of the genus *Paramecium* and multicellular forms. Somewhat more variable is a domain of the large subunit, and the two internal transcribed spacers (ITS1 and ITS2), on either side of the 5.8s RNA. These regions have also been used in studies of evolution and are more favorable for somewhat closer relatives than the 16s rDNA. Colman[6]

Ribosomal RNA and DNA

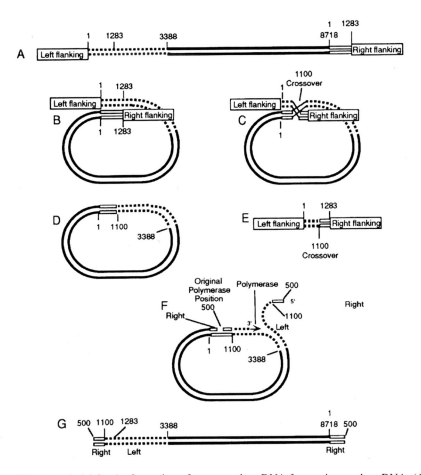

FIGURE 9.2 Model for the formation of macronuclear DNA from micronuclear DNA. (A) Linear micronuclear RIB2. (B) The molecule bends back on itself and homologous regions pair. (C) Next there is a crossover at about base 1100, which cuts out the region with the deletion of the micronuclear gene. (D) Resultant circular form of the micronuclear gene. (E) Defective region that was cut out. (F) A polymerase molecule now attaches at about base 500 and begins to transcribe in a typical rolling-circle replication. The left part of the molecule peeling off of the template begins at about base 500, and the polymerase continues to transcribe, going around the circle to the position where replication started. (G) Molecule produced by one cycle of replication. The polymerase continues around the circle numerous times to produce a continuous sequence of transcribed and nontranscribed regions. Finally, replication stops near the point where it started, producing the right end of the molecule. After a complementary strand has been synthesized, telomeres are then added to the free ends of the linear molecule to produce the macronuclear polymer of RIBX. Alternatively, the two free ends, instead of adding telomeres, may combine to form a circle. With permission from the American Society for Microbiology; from Preer, L. B., et al., *Mol. Cell. Biol.*, 19, 7792, 1999, Figure 10.

sequenced the internal transcribed spacer regions of thirteen of the fourteen species of the *P. aurelia* complex and used them in an evolutionary study of species within the complex. She also established folding patterns for these sequences. She concluded that the variation from species to species within the sibling species of the *P. aurelia* complex is about like that seen in other groups of sibling species, such as those in *Drosophila*. A variable domain of the large subunit was used by Nanney et al.[7] in studying evolutionary relations between *Tetrahymena* and *Paramecium*. They found a large amount of variation from species to species in the *P. aurelia* complex and suggested that these species had been separated for a very long time. Variations in a domain of the large subunit and the internal transcribed spacer region (ITS1) among species 12, 13, and 14 in the different species within the *P. aurelia* complex were noted by Tarcz et al.[8] These sequences were also used in studying the evolutionary relationships among the different forms. The most variations were found in the species 12 group, which they suggest, on the basis of their results, might be made up of several additional subspecies.

REFERENCES

1. Findly, R. C. and Gall, J. G. 1978. Free ribosomal RNA genes in *Paramecium* are tandemly repeated. *Proc. Natl. Acad. Sci. U.S.A.* 75:3312.
2. Findly, R. C. and Gall, J. G. 1980. Organization of ribosomal genes in *Paramecium tetraurelia*. *J. Cell Biol.* 84:457.
3. Phan, H. L., Forney, J., and Blackburn, E. H. 1989. Analysis of *Paramecium* macronuclear DNA using pulsed field gel electrophoresis. *J. Protozool.* 36:402.
4. Preer, L. B., et al. 1999. Does ribosomal DNA get out of the micronuclear chromosome in *Paramecium tetraurelia* by means of a rolling circle? *Mol. Cell. Biol.* 19:7792.
5. Sogin, M. L. and Elwood, H. J. 1986. Primary structure of the *Paramecium tetraurelia* small sub-unit rRNA coding region: Phylogenetic relationships within the *Ciliophora*. *J. Mol. Evol.* 23:53.
6. Colman, A. W. 2005. *Paramecium aurelia* revisited. *J. Eukar. Microbiol.* 52:68.
7. Nanney, D. L., et al. 1998. Comparison of sequence differences in a variable 23S rRNA domain among sets of cryptic species of ciliated protozoa. *J. Eukary. Microbiol.* 45:91.
8. Tarcz, S., et al. 2006. Intraspecific variation of diagnostic rDNA genes in *Paramecium dodecaurelia*, *P. tredecaurelia* and *P. quadecaurelia* (Ciliophora: Oligohymenophorea). *Acta Protozool.* 45:255.

10 Cortical Morphogenesis

10.1 DOUBLE ANIMALS

Tracy Sonneborn started his career in biology studying the flatworm, *Stenostomum*.[1] He found that when he treated the worms with lead acetate he was able to induce individuals with two heads, rather than the normal one. This trait proved to be stable at reproduction. Although he suspected that the trait was inherited cytoplasmically, he was unable to determine decisively, for no sexual reproduction was available.

Somewhat later he studied the ciliate *Colpidium*.[2] Here he found that if the cells were cultured on a specific strain of bacterium they formed chains, which produced double cells after further reproduction. These doubles were also stable. Again, he was unable to make crosses because of the lack of sexual reproduction, and again he was unable to analyze their inheritance. However, based on the circumstances of their origin, it was likely that something other than gene mutation was responsible for the double cells.

Much later in his career, Sonneborn[3] visited the phenomenon of double animals again. By this time, however, he had developed sophisticated methods of genetic analysis with the *P. aurelia* group in regard to both genically and cytoplasmically controlled characters. Doublet cells were obtained in *Paramecium* by ensuring that the cells were both the same serotype and inducing them with a brief exposure to homologous antiserum. A cross of a line of doublets to a line of singles produced doublets from the double parent and singles from the single parent. No change occurred when F1 was carried through subsequent autogamy. So the trait was clearly not due to genic differences.

If the trait was not genic he suspected that it might be cytoplasmic, so he marked the cytoplasm of one or the other of the two forms with kappa and made the cross again, ensuring that mixture of the two cytoplasms took place. This was done in the cross of doublets with singlets, one of them having kappa, the other not. Now he made sure that both of the cultures to be crossed were of the same serotype and then exposed the conjugants to homologous antiserum, producing delayed separation and the mixing of cytoplasms so that both exconjugants now had kappa. It was again found that exconjugants derived from doublet conjugants yielded doublets, and exconjugants derived from singlets yielded singlets. So differences in the fluid cytoplasm were not the basis of singles and doubles.

Sonneborn also eliminated the possibility that the trait was determined by the macronuclei, that is, macronuclear or caryonidal inheritance. He did this by making the cross and inducing macronuclear regeneration. Examination of the phenotypes of the individuals segregating after conjugation established that no fragments of the old macronuclei were present, and thus could not be responsible for the inheritance of doublets. He concluded that the cortex of the cell contained the determinants for double animals.

This novel hypothesis was then tested by finding that at conjugation cells sometimes have a portion of the cortex torn off one mating cell and implanted in the other. Such a process sometimes occurs spontaneously if delayed separation of conjugants occurs or is induced by exposure to antiserum during conjugation. In this way, it was possible to implant an extra piece of cortex, bearing certain readily identifiable structures, into recipient cells that subsequently produce new clones bearing the same pieces of the cortex. Such modifications in these cells were also inherited like double animals.[3]

10.2 CORTICAL MUTANTS

The next advance in our understanding of the cortex[4] requires an understanding of the cortical structure. It should be recalled that the surface of *Paramecium* is made up of kineties, rows of unit structures containing cilia and associated structures. See Chapter 3. The structures within each kinety are arranged asymmetrically, and the kineties have a polarity, running from the front of the cell to the back. At conjugation it was found that during serum-induced delayed separation, sometimes the conjugants of a pair are found to be reversed. After separation such abnormal pairs sometimes produce progeny in which a part of the cortex of one of the cells appears to have been torn off and inserted into the other conjugant in a reverse orientation. See Figure 10.1. Cells carrying the transplanted piece of cortex consist of only a few kineties (up to twelve), and the cells are said to have inverted kineties. They are called reversed polarity (RP) mutants. If the surface of the cell is silver stained, the inverted kineties are easily recognized and are inherited as a cortical structure in the same way that double animals are inherited; even the number of kineties in a particular mutant is maintained. RP mutants pursue a "twisty" course when swimming forward, because the cilia in the inverted kineties beat in the opposite direction from those of the normally oriented kineties of the cell. Such rows of reversed units constitute the first reported exception to the rule of desmodexy—that kinetodesmal fibers in ciliates always lie to the right of the kinetosomes when observed from the ventral surface of the cell.[5]

Sonneborn[6] pointed out that the kinetosomes commonly arise so close to preexisting kinetosomes that it had not been surprising that Lwoff (p. 5)[7] had earlier described the process as one of division. Lwoff's observations were made, of course, before the electron microscope was in general use.

To interpret these findings, it is necessary to take into consideration the growth of individual surface organelles, especially the kinetosomes (ciliary basal granules). Detailed electron microscope studies by Dippell,[8] working in Sonneborn's laboratory, showed that before replication each unit within the kinety consists of a kinetosome, a kinetodesmal fiber to the left (when viewed from the dorsal surface of the cell) and running forward, a parasomal sac to its left, and a trichocyst insertion point between units. See Figure 3.2. At reproduction, below the surface of the cell, a new kinetosome is produced in front of and at right angles to the old one. It then moves up, rotates on its axis, and comes to lie with the same orientation as the old kinetosome, in front of the old one. A new kinetodesmal fiber appears and runs to the left and forward like the old, and a new parasomal sac is created to the left. Now the unit

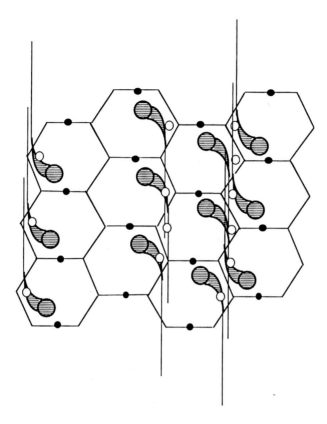

FIGURE 10.1 Diagram of an inverted kinety mutant. The two kineties on the far left and far right are normal (see Figure 3.2), while the central two kineties are inverted. The structures within each unit have normal relations to each other. Inverted kineties are easily recognized microscopically.

simply divides into two and a new trichocyst is inserted, making the kinety longer by one unit. Dippell also showed that if a kinety can be reversed, the same division process is repeated but in reverse, such that the new unit fits in with the preexisting units of the reversed kinety, and ensures that the reversed kinety is inherited. These observations have been confirmed by several other workers.[9-11]

It should be added that Sonneborn[6] was unable to find any evidence that the kinetosomes of *Paramecium* contain DNA, as had been claimed by previous workers.[12]

Sonneborn summed up his work on the cortex by asking the question: What are the determinative molecular species, configurations, and modalities in the cortex? In his view, the answer remained almost completely unknown, not only in ciliates, but in all organisms. There was no reason to doubt that the molecules making up the cortex were coded in DNA, or formed by the activity of DNA-coded enzymes, since every feature of the cortex could be altered by gene mutations (as listed in Sonneborn's 1975[13] review). The basis for this kind of inheritance lay, apparently, in the organizing and developmental function of the existing cortical structure, both visible and molecular, and was referred to by Sonneborn as cytotaxis (p. 925).[14] The products of genic action could not initiate this cortical structure in the absence of an existing guiding pattern of development. Sonneborn therefore saw no escape from the revolutionary conclusion that certain aspects of development in *Paramecium*, and presumably in other organisms too, were encoded partly in cortical geography and not solely in DNA.

For more extended views of cortical inheritance and its relation to organisms other than *Paramecium*, see books by Frankel[15] and Grimes and Aufderheide.[16] Frankel states that the ease of detection of inheritance of structural systems in ciliates is a consequence of ciliate topology rather than any particular mechanism of propagation: any structural system can be inherited in the ciliate clonal cylinder if it can grow longitudinally and maintain lateral boundaries while it is growing. These considerations provide the reason why Frankel prefers not to use any single term, such as Sonneborn's *cytotaxis*. He does, however, use the term *structural inheritance*, which emphasizes the likelihood that we are dealing with information at a supramolecular level, but that has as few mechanistic connotations as possible.

It seems to us that Ruth Dippell's discoveries provide us with some understanding of cortical inheritance at an anatomical level. Studies at the next level must deal with the molecular aspects of why structures are formed next to adjacent structures and what determines the orientation of new structures in relation to preexisting structures when they are formed. These seem to be very difficult problems indeed, for many substances are involved in the production of new kinetosomes. Nevertheless, a beginning is being made. Tubulins are being studied by Ruiz et al.[17] and Garreau de Loubresse et al.[18] Work is also being done on centrins.[19-22] Others are looking at the role of protein phosphorylation.[23,24] To gain molecular information that explains cortical inheritance, however, will be difficult.

10.3 TRICHOCYSTS

We now turn our attention to one of the best-studied structures in *Paramecium*, the trichocysts. The trichocysts are sac-like organelles, situated just beneath the cell surface of *Paramecium*, oriented perpendicularly to the cell surface, and precisely inserted at sites in the cross-ridges between the cilia-bearing units. They are protein crystals,[25] both before and after extrusion. See Figures 3.9 and 3.10. When the cell is stimulated with various chemicals (e.g., picric acid), the trichocysts are explosively discharged into the medium, forming long needle-like structures. As discussed in Chapter 3, the function of the trichocysts is protection against various protozoan predators. Since the trichocysts liberate their contents into the outside medium they are often called extrusible or secretory organelles.

After isolation and solubilization of the trichocysts, their proteins can be run on two-dimensional gels. The result is some thirty-five major protein spots, and counting minor spots, up to one hundred total.[26,27] Investigations show that the proteins are coded by separate genes. The one hundred proteins make a gigantic protein crystal, which changes its form when discharged.

Another unique feature of trichocysts is their formation deep within the cytoplasm, transport to the cell surface by saltation,[28] and then attachment at a specific site on the membrane surface.[29,30] The area is distinguished by numerous granules in distinctive positions, the most characteristic of which is a ring of granules marking the trichocyst attachment site. Trichocysts must be attached in order to be extruded.

Numerous single gene mutations affecting trichocysts have been induced.[13,31] All of the mutants are incapable of discharging their trichocysts, and many have morphological defects as well. Pollack[31] gave the morphological mutants descriptive names reflecting their form (football, pointless, cigar), and in the absence of trichocysts altogether, the mutant was called trichless. Most are nondischarge (ND) mutants. Some are temperature sensitive (normal at a low temperature, mutant at a higher temperature).

Aufderheide[32] found that when a few (four to six) undischarged trichocysts from some mutants (unable to discharge, but with normal-appearing trichocysts) are injected into the football mutant of Pollack, the individual injected trichocysts respond by discharging the injected trichocysts when challenged with picric acid. These few discharged trichocysts could readily be seen protruding from the surface of the cell when observed with a microscope at 100× magnification. It is assumed that the reason these mutant trichocysts can be discharged in their new environment is that they have good proteins in their trichocysts, but are deficient in their cytoplasms or cell membrane attachment regions. These deficiencies are corrected by the cytoplasms and membranes of the cells into which they are transferred, in this case into the mutant "football," whose defect lies only in the trichocyst protein. On the other hand, the transferred trichocysts of other nondischarge mutants are not able to respond when injected in a similar fashion because their trichocyst proteins are defective and are not corrected by the cytoplasm and membranes of the cells into which they are injected. Examination of the mutants deficient in their cytoplasms could be distinguished from mutants deficient in their membrane attachment points by examination with freeze-fracture electron microscopy. Thus, it possible to distinguish all three sites of mutation: trichocyst proteins, cytoplasm, and membrane attachment regions.[32-35]

Instead of transferring trichocysts by microinjection, one can simply cross mutants to provide a test similar to that described above. A mutant in question is crossed to the football trichocyst mutant of Pollack, and then the exconjugant from the mutant side of the cross is challenged with picric acid after the paramecia have been together 3 or 4 hours and before the nuclei have exchanged. Just placing the two cell membranes in close association will allow exchange of soluble constituents, and also allow discharge of trichocysts if the defect is in the cytosol or in the membranes themselves. The results of the microinjection and of the conjugation tests are in agreement.

Current studies deal with very general mechanisms found to influence secretory processes in all organisms. These include work on N-ethylmaleimide-sensitive factor (NSF), a protein involved in membrane fusion, and protein complexes called soluble NSF attachment receptors (SNAREs), which are important in membrane fusion.[36,37] Proton ATPases are found to be important in a study by Wassmer et al.[38] Studies are being carried out on cyclic guanine monophosphate (GMP) protein kinases, which phosphorylate proteins.[39] Work is also being done on a protein called regulator of chromosome condensation (RCC).[35] Finally, there is an investigation of the role of Ca^{++} in trichocyst discharge.[40]

While most trichocyst mutants are inherited as single gene mutations, a few are inherited cytoplasmically and will be discussed in Chapter 12.

REFERENCES

1. Sonneborn, T. M. 1930. Genetic studies on *Stenostomum incaudatum* (nov. spec.). *J. Exp. Zool.* 57:409.
2. Sonneborn, T. M. 1932. Experimental production of chains and its genetic consequences in the ciliate *Colpidium campylum*. *Biol. Bull.* 63:187.
3. Sonneborn, T. M. 1963. Does preformed cell structure play an essential role in cell heredity? In *The nature of biological diversity*, ed. J. M. Allen, 165. New York: McGraw Hill.
4. Beisson, J. and Sonneborn, T. M. 1965. Cytoplasmic inheritance of organization of the cell cortex in *Paramecium aurelia*. *Proc. Natl. Acad. Sci. U.S.A.* 53:275.
5. Chatton, E. and Lwoff, A. 1930. Imprégnation, par diffusion argentique de l'infraciliature des cilieé marins et d'eau douce, après fixation et sans dessication. *Compt. Rend. Soc. Biol.* 104:834.
6. Sonneborn, T. M. 1970. Gene action in development. *Proc. R. Soc. Lond.* 176B:347.
7. Lwoff, A. 1950. *Problems of morphogenesis in ciliates: The kinetosomes in development, reproduction and evolution.* New York: John Wiley & Sons.
8. Dippell, R. V. 1968. The development of basal bodies in *Paramecium*. *Proc. Natl. Acad. Sci. U.S.A.* 61:461.
9. Iftode, F., et al. 1989. Development of surface pattern during division in *Paramecium*. I. Mapping of duplication and reorganization of cortical cytoskeletal structures in the wild type. *Development* 105:191.
10. Aufderheide, K. J., Rotolo, T. C., and Grimes, G. W. 1999. Analyses of inverted ciliary rows in *Paramecium*. Combined light and electron microscopic observations. *Eur. J. Protistol.* 35:81.
11. Iftode, F. and Fleury-Aubusson, A. 2003. Structural inheritance in *Paramecium*: Ultrastructural evidence for basal body and associated rootlets polarity transmission through binary fission. *Biol. Cell* 95:39.
12. Smith-Sonneborn, J. and Plaut, W. 1967. Evidence for the presence of DNA in the pellicle of *Paramecium*. *J. Cell Sci.* 2:225.
13. Sonneborn, T. M. 1975. *Paramecium aurelia*. In *Handbook of genetics*, ed. R. C. King, chap. 20. Vol. 2. New York: Plenum Press.
14. Sonneborn, T. M. 1964. The differentiation of cells. *Proc. Natl. Acad. Sci. U.S.A.* 51:915.
15. Frankel, J. 1989. *Pattern formation: Ciliate studies and models.* New York: Oxford University Press.
16. Grimes, G. W. and Aufderheide, K. J. 1991. *Cellular aspects of pattern formation: The problem of assembly*, ed. H. W. Saur. Monographs in Developmental Biology 22. New York: Karger.
17. Ruiz, F., et al. 1999. Basal body duplication in *Paramecium* requires gamma-tubulin. *Curr. Biol.* 9:43.
18. Garreau de Loubresse, N. G., et al. 2001. Role of delta-tubulin and the C-tubule in assembly of *Paramecium* basal bodies. *BMC Cell Biol.* 2:4.
19. Madeddu, L., et al. 1996. Characterization of centrin genes in *Paramecium*. *Eur. J. Biochem.* 238:121.
20. Klotz, C., et al. 1997. Genetic evidence for a role of centrin-associated proteins in the organization and dynamics of the infraciliary lattice in *Paramecium*. *Cell Motil. Cytoskel.* 38:172.

21. Vayssié, L., Sperling, L., and Madeddu, L. 1997. Characterization of multigene families in the micronuclear genome of *Paramecium tetraurelia* reveals a germline specific sequence in an intron of a centrin gene. *Nucl. Acids Res.* 25:1036.
22. Ruiz, F., et al. 2005. Centrin deficiency in *Paramecium* affects the geometry of basal body duplication. *Curr. Biol.* 15:2097.
23. Keryer, G., et al. 1987. Protein phosphorylation and dynamics of cytoskeleton structures associated with basal bodies in *Paramecium*. *Cell Motil. Cytoskel.* 8:44.
24. Sperling, L., et al. 1991. Cortical morphogenesis in *Paramecium*: A transcellular wave of protein phosphorylation involved in ciliary rootlet disassembly. *Dev. Biol.* 148:205.
25. Sperling, L., Tardieu, A., and Gulik-Krzywicki, T. 1987. The crystal lattice of *Paramecium* trichocysts before and after exocytosis by x-ray diffraction and freeze-fracture electron microscopy. *J. Cell Biol.* 105:1649.
26. Shih, S. J. and Nelson, D. L. 1991. Multiple families of proteins in the secretory granules of *Paramecium tetraurelia*: Immunological characterization and immunocytochemical localization of trichocyst proteins. *J. Cell Sci.* 100:85.
27. Madeddu, L., et al. 1995. A large multigene family codes for the polypeptides of the crystalline trichocyst matrix in *Paramecium*. *Mol. Biol. Cell* 6:649.
28. Aufderheide, K. J. 1978. Genetic aspects of intracellular motility: Cortical localization and insertion of trichocysts in *Paramecium tetraurelia*. *J. Cell Sci.* 31:259.
29. Plattner, H., Miller, F., and Bachmann, L. 1973. Membrane specializations in the form of regular membrane-to-membrane attachment sites in *Paramecium*. A correlated freeze-etching and ultrathin sectioning analysis. *J. Cell Sci.* 13:687.
30. Beisson, J., et al. 1976. Genetic analysis of membrane differentiation in *Paramecium*. Freeze-fracture study of the trichocyst cycle in wild-type and mutant strains. *J. Cell Biol.* 69:126.
31. Pollack, S. 1974. Mutations affecting the trichocysts in *Paramecium aurelia*: Morphology and description of the mutants. *J. Protozool.* 21:352.
32. Aufderheide, K. J. 1978. The effective site of some mutations affecting exocytosis in *Paramecium tetraurelia*. *Mol. Gen. Genet.* 165:199.
33. Lefort-Tran, M., et al. 1981. Control of exocytotic processes: Cytological and physiological studies of trichocyst mutants in *Paramecium tetraurelia*. *J. Cell Biol.* 88:301.
34. Vayssié, L., et al. 2000. Molecular genetics of regulated secretion in *Paramecium*. *Biochemie* 82:269.
35. Gogendeau, D., et al. 2005. Nd6p, a novel protein with RCC1-like domains involved in exocytosis in *Paramecium tetraurelia*. *Eukaryot. Cell* 4:2129.
36. Froissard, M., et al. 2002. N-ethylmaleimide-sensitive factor is required to organize functional exocytotic microdomains in *Paramecium*. *Gen. Soc. Am.* 161:643.
37. Kissmehl, R., et al. 2002. NSF regulates membrane traffic along multiple pathways in *Paramecium*. *J. Cell Sci.* 115:3935.
38. Wassmer, T., et al. 2005. The vacuolar proton-ATPase plays a major role in several membrane-bounded organelles in *Paramecium*. *J. Cell Sci.* 118:2813.
39. Kissmehl, R., et al. 2006. Multigene family encoding 3',5'-cyclic GMP-dependent protein kinases in *Paramecium tetraurelia* cells. *Eukaryot. Cell* 5:77.
40. Mohamed, L., et al. 2002. Functional and fluorochrome analysis of an exocytotic mutant yields evidence of store-operated Ca^{++} influx in *Paramecium*. *J. Membrane Biol.* 187:1.

11 Behavior

11.1 INTRODUCTION

One of the first students of behavior in *Paramecium* was H. S. Jennings, who published his studies in a book called the *Behavior of the Lower Organisms*.[1] Jennings described in great detail the normal swimming behavior of *Paramecium*, including one of the major behavioral responses, the avoiding reaction, in which *Paramecium* at times would be stimulated, for example, by touching its anterior end on an obstacle in the medium. It then would stop its forward progress, reverse its direction of swimming, and then resume forward progress in a different direction. Other behaviors—thigmotaxis, geotaxis, chemotaxis, thermotaxis, and galavanotaxis—were also noted by Jennings.

11.2 MUTANTS

Work on behavioral mutants in *Paramecium* was continued by Kung,[2,3] who found numerous mutants in *Paramecium tetraurelia*. He gave these new mutants appropriate and imaginative names that reflected the character of the mutant. Thus, a mutant named "paranoiac" gave a continuous avoiding reaction when disturbed; "pawn" never gave avoiding reactions and was only capable of swimming forward (named for the chess piece that can only move forward); "atalanta," after the Greek story, swam in a jerky fashion (as if stopping to pick up golden apples); "fast" swam very fast; "staccato" had frequent avoiding reactions; "spinner" tended simply to stop and spin in place upon reaching the edge of the dish or when properly stimulated; and so on.

11.3 GENETICS

Crosses of the different mutants to wild type and to each other[4] reveal their genetic relationships. Both multiple genes and multiple alleles were found. Numerous loci have been discovered. Some alleles were dominant and some recessive. There are also cases of cytoplasmic inheritance of behavioral mutants, and these will be discussed in Chapter 12.

11.4 CELL MODELS

In trying to understand the nature of the mutants, Kung quickly became aware that the key was to be found in the membranes of the cell. Thus, cell models can be prepared by treating paramecia with Triton X-100, which kills the cells and removes their membranes. In the case of pawn, which can only swim forward, pawn models, that is, pawn cells, like wild type, treated with Triton-X and lacking membranes, although dead, can be made to swim forward if one adds adenosine triphosphate (ATP) and Mg^{2+} or Ca^{2+}. And if even more Ca^{2+} is added, then swimming is backward. Thus, the inability of pawn to swim backward apparently lies in a defect in its membranes.[5,6]

It is clear that increases in the internal Ca^{2+} concentration lead to reversed swimming, but it is unknown by what mechanism this comes about in *Paramecium*. Why should change in the concentration of various ions within the cell, or just change in the electrical properties of the cell membrane, lead to changes in the speed or direction of beating of cilia? In higher organisms in muscle contraction, the relation between the stimulus and the reaction within the cell has been studied in great detail and has resulted in the well-known sliding filament model of Huxley. It depends on the interaction of numerous elements such as Ca^{2+}, myosin, ATP, and so on. In *Paramecium* a sliding filament model (the filaments are different in cilia) has also been proposed.[7] However, we still do not know the details of how these stimuli produce changes in ciliary beating.

11.5 MEASURING THE ELECTRIC PROPERTIES OF THE CELL MEMBRANE

It has long been known that the membranes of higher organisms show action potentials, and the situation in muscle systems has long been investigated. So when it was discovered in the 1950s that the membranes of *Paramecium* could produce action potentials, the finding was greeted with great interest by the biological community. It was hoped that elucidation of mechanisms in the lower organisms would lead to a better understanding of the mechanisms and the evolution of nervous systems in general. Indeed, as we shall see, studies on *Paramecium* have produced concepts that have advanced our knowledge of ion channels.

How can the properties of the membranes be measured? At first, only single electrodes were used, one inside the cell and one outside. Stimulation of wild type to produce an avoiding reaction produced an extremely rapid but complicated response, measured in milliseconds, consisting of a series of spikes in voltage that gradually decreased in amplitude over a few milliseconds.[4]

While the lipid bilayer that makes up the cell membrane is impermeable to various positive ions, it turns out that there are numerous ion channels, made up of one or more molecules, that can open and allow the passage of ions into and out of the cell. Often the opening of one channel induces the opening of another channel, either indirectly by increasing another ion in the cell, or directly by a change in voltage. One of the ways of studying these channels is, after adjusting the ions to which the cells are exposed, to place one electrode within the cell, another outside, and a third inside to apply a voltage differential to the inside and outside of the cell. With this arrangement one may set the voltage to any desired value, producing a "voltage clamp." Now a momentary stimulus (achieved by a quick change in the voltage, followed by restoration of the voltage adjusted to normal) results in the opening of channels and the movement of positively charged ions, Ca^{2+}, K^+, Na^+, and so on, in and out. These movements produce temporary differences in electrical currents on the order of milliseconds that can be measured.[8] The voltage clamp was invented by Kenneth Cole in the 1940s, and the interpretation of the results can be quite complex.

Today studies on behavior are tied to a study of individual channels, using a technique called the patch clamp. The technique was developed by Erwin Neher and Bert Sakmann in the late 1970s and early 1980s, for which they received the Nobel Prize in Physiology and Medicine in 1991. The lipid bilayer that makes up the membranes of a cell is impermeable to positive ions. But since we know that many ions can move through membranes, we know that channels must exist, each with a different set of properties. It has been found that a 1-micron^2 area of a cell is big enough to hold a single channel. Therefore, a micropipette with a 1-micron-diameter tip can be pushed up to a membrane, and if gentle suction is now applied, a patch of the cell membrane will be trapped across the opening of the tip. It has been found that often a single channel will be obtained in the patch. Now with the proper electronic equipment, if electrical contact is made between this tip and the medium, the voltage may be clamped to the resting voltage of the membrane and variations may be measured after changes in voltage, or the introduction of new ions into the medium. Only two electrodes are required. Membranes may be attached in the normal way, original outside of the membrane to the outside of the electrode tip, or in the reverse way, or "inside out" orientation. More recent studies on membrane channels have used this technique, for much clearer results are obtain with patch clamps.[9,10]

11.6 THE AVOIDING REACTION

What happens in a typical avoiding reaction? Kung and Saimi[11] and Ramanathan et al.[12] describe a series of steps. Although the avoiding reaction can be induced by touch, Kung found that the avoiding reaction that Jennings described could be induced by other stimuli. Thus, if wild-type cells are allowed to sit in a medium devoid of Na$^+$ (the adaptation solution, 4 mM KCl, 1 mM CaCl$_2$) and then transferred to a medium containing Na$^+$ (the testing solution, 4 mM NaCl, 1 mM CaCl$_2$), avoiding reactions are induced. (These solutions also contain a constant amount of Tris buffer and citrate.) A touch on the anterior end, or a change in the ions in the medium, or a temporary change in voltage creates a change in potential across the cellular membrane called a receptor potential. The opening and closing of the channels is monitored by a voltage clamp, which is used to maintain a given voltage. The temporary change in voltage now very quickly opens a Ca^{2+} channel, letting Ca^{2+} run *into* the cell (inward inactivation current), and later another channel, the K$^+$ channel, is also stimulated to open by the increasing concentration of Ca^{2+}, allowing K$^+$ to flow through the membrane *to the outside* (delayed rectification current). The Ca^{2+} change is measured in microseconds, while the K$^+$ change is slower and is measured in milliseconds. If the external medium contains Na$^+$ ions, then a Ca^{2+}-activated inward Na$^+$ current develops. The increasing Ca^{2+} also induces a reversal of the direction of the ciliary beat, causing the cell to back up. After the reactions are finished, the original resting potentials are restored and the cell swims forward again, usually in a new direction.

The parallel between the avoiding reaction and the processes that occur in higher organisms is striking. We see that a stimulus in *Paramecium* opens Ca^{2+} channels to depolarize the cell membrane, and repolarization occurs after the delayed opening of K+ channels. However, in higher organisms, when a neuron is stimulated it opens Na+ channels to depolarize the cell membrane and (like in *Paramecium*) is restored by the opening of K+ channels. These were very exciting results when first discovered, and led to increased interest and study of the membranes of *Paramecium*.

11.7 ION CHANNELS

Mutants have played a major role in discovering and identifying the ion channels. Mutants generally have only a single function missing, and examination of their activities with a voltage clamp or patch clamp characterizes this function. Most of the mutants, when injected with the wild-type genes, are able to function normally. This fact has led to the isolation of the affected locus in many cases, for example, the locus that codes for the protein calmodulin. A few of the mutants and ion channels are listed briefly in the following paragraphs. Most of the references on which these results are based are found in the more detailed papers by Ramanathan et al.[12] and Saimi and Kung,[13] to which the reader is referred.

11.7.1 CA^{2+} CHANNELS

The *cnr* (ciliary nonreversible) mutants, *cnrA*, *cnrB*, *cnrC*, and *cnrD*, of *P. caudatum* and the "pawn" mutants, *pwA*, *pwB*, and *pwC*, of *P. tetraurelia* have no Ca^{2+} channels and produce results consistent with the absence of these channels in voltage clamps on the mutants. Presumably, the wild-type allele of each one of these mutant genes codes for a protein that is essential for the channel, and each mutant lacks one of the essential proteins. The "dancer" mutation, *Dn*, has a stronger Ca^{2+} ion current than normal. The wild-type gene of the dancer mutation has been cloned by functional complementation.[14] See Chapter 4 for a discussion of functional complementation. This is the first gene cloned using this technique. The open reading frame (ORF) is about 500 bp long and contains a 26 bp intron that is removed after mRNA is made. It appears to code for a novel protein.

The pawn B gene has also been cloned by functional complementation.[15] It rescues pawn B, but not pawn A, when injected. The pawn B ORF is 984 bp in length, and its product matches no other known protein. Two of the allelic mutants prove to have single base substitutions in the gene, and another represents a mutation in a 44 bp IES (IES 427) that prevents its excision. Another IES (IES 658) was also found in the gene; the last nine bases of this IES are often not excised and represent an alternative splice site. The gene was also found to have a 29 bp intron.

At least three ion channels are active in passing Ca^{2+} through the cell membrane. They are ones activated by touch[16] or depolarization.[17–19]

11.7.2 K+ CHANNELS

"Pantophobiac" mutants prevent the opening of Ca^{2+}-activated K+ channels. These mutants give strong backward swimming in a variety of solutions. The mutant *Tea* (tetraethylammonium) insensitive has Ca^{2+}-activated K+ channels that are activated more easily than the same channels in wild type.

Injections of cellular fractions from wild type into the pantophobiac mutant are able to cure the swimming defects in the mutant.[12,20] Moreover, when the active factor was isolated, it turned out to be calmodulin, a well-known protein known to bind Ca^{2+} and important in numerous Ca^{2+} activities in the cell. When calmodulin was injected into the mutant, it was highly active in curing the mutant. It has been shown that *Paramecium* calmodulin can be expressed in *E. coli* if changes are made in the gene to make its genetic code compatible with the universal code.[21] Numerous genes make up the family of proteins able to combine with calmodulin, reflecting the numerous functions of calmodulin in *Paramecium*.[22] It is thought that one of the functions of calmodulin from *Paramecium* is to serve as the protein making up the Ca^{2+} ion channel.[23]

Aside from these genes, there are other K^+ channels in other organisms whose sequences are known. Probing the genomic *Paramecium tetraurelia* library with these sequences has led to the discovery of a large family of genes, the *PAK* (*Paramecium* K^+) channel genes.[24] If an isolated gene is injected into a wild-type *Paramecium*, gene silencing results, and the resulting phenotype is like that of the pantophobiac mutant.[25] In 2001 the family consisted of at least several additional K^+ channel genes, with many more likely to exist. Several had been characterized. One was activated by depolarization;[26,27] a second slower one was activated by the entrance of Ca^{2+} from depolarization;[28] a third one was activated by hyperpolarization;[29] a fourth was activated by the entrance of Ca^{2+} from hyperpolarization;[30] and a fifth was activated by touch at the posterior end of the cell.[31] Presumably, there are numerous others activated by subtle differences in the environment.

11.7.3 Na+ Channels

The "fast mutant," *fna-2*, does not show avoiding reactions when placed into a medium containing Na^+, as does wild type. The Ca^{2+}-activated Na^+ channel is blocked.[32] On the other hand, paranoiacs enhance the Ca^{2+}-activated Na^+ currents. Saimi and Ling,[33] using excised membrane patches, showed that calmodulin could activate the Ca^{2+}-dependent Na^+ channels, possibly by reacting with the Ca^{2+} channel protein.

11.7.4 Mg2+ Channels

The mutant *eccentric* fails to show avoiding reactions in a medium containing Mg^{2+}. Moreover, it has been shown that wild type has an efficient Mg^{2+} ion channel. The gene for this protein was isolated using functional complementation. However, the protein appears to be more closely related to exchangers (ions that simply exchange ions for other ions on the surface of the cell) than to ion channel proteins, but the relationship between the two is somewhat vague at present.[25] The *Cha* (chameleon) mutants are also involved in Mg^{2+} conductance.[34]

11.8 BEHAVIORAL RESPONSES OTHER THAN THE AVOIDING REACTION

Behavioral responses other than the avoiding reaction occur in wild type. They include response to touch, gravity, light, temperature, attractants in the medium, and electrical charge. These are all thought to be manifested by the opening and closing of ion channels different from the ones seen above in the avoiding reaction. For a detailed consideration of these responses, the reader is referred to the numerous papers by Machemer, reviewed in Machemer and Teunis[35] and Van Houten.[36]

When *Paramecium* is touched on the rear, it speeds up its forward motion. Thus, Bonini et al.[37] and Van Houten et al.[38] postulate that a touch on the rear of a cell causes a hyperpolarization, which produces an increase of cyclic adenine monophosphate (cAMP), followed by protein phosphorylation and an increase in ciliary beat.

Mutants that depress the sensitivity of certain chemical attractants can sometimes be found. A mutant discovered by Van Houten[39] is repelled by Na acetate, which attracts wild type. The well-known G-proteins (guanidine binding) act as tranducers of exterior stimuli. Inhibitors of G-proteins inhibit attraction of wild-type *Paramecium* to Na acetate and ammonium chloride. It is also found that Ca^{2+} influx is inhibited by interference with G-proteins.[40]

Identifying these channel proteins has posed a problem for investigators. Paquette et al.[41] note that surface proteins in *Paramecium* are often fixed into the membrane by lipid glycophosphatidylinositol (GPI) anchors (originally shown for the very large immobilization antigens by Capdeville et al.[42]). By removing the anchors, by using either an exogenous enzyme or salt-alcohol to activate an enzyme already present in *Paramecium*, both the group of very large GPI-anchored immobilization antigens and a number of smaller GPI-anchored proteins are released. Antibodies prepared against many of these smaller proteins often produce behavioral defects. These genes can then be isolated and studied, and thus provide a different way of obtaining many of the desired receptor proteins. The protein necessary for the response to folate was isolated in this way.

It has also been possible to identify in *Paramecium* a gene with sequences like the GPI-synthesizing enzyme, *pPIG-A*, known from other organisms. Injection of the *Paramecium* gene, or homologues, into wild type can lead to gene silencing (see Chapter 12), and strains containing the silenced gene have modified behavior in response to chemicals, providing additional evidence that proteins that cause chemical behavioral responses are also GPI anchored.[43]

11.9 CONCLUSIONS

Due to the work on *Paramecium* in the 1970s and later, our concept of the ion channel began to change. Although the initial work led to the isolation of some of the proteins of ion channels, the core proteins, later experiments began to complicate that initial simple picture. Thus, when pantophobiac was injected with an extract of wild-type *Paramecium*, it was found that pantophobiac was temporarily "cured" of its abnormal behavior. One might have expected the active element in the extract to be the core protein. As already noted, it proved to be a well-known soluble protein, calmodulin. Other peculiarities were also found. The mutant gene in pantophobiac (affecting a Ca^{2+}-activated K^+ channel) was found to be allelelic to the mutant gene in fast-2 (affecting a Ca^{2+}-activated Na^+ channel), and both were cured by calmodulin. When the two genes were cloned, they proved to control variants in two parts of the calmodulin molecule. Furthermore, patch clamp studies have shown that when calmodulin is added to the medium, it can activate or deactivate the channel. The final conclusion is that ion channels are more complex than we had originally thought. They consist of a pore protein and other proteins such as calmodulin that affect the functioning of the pore.

Paramecium has an amazing number of genes involved in controlling its behavior. Note that *Paramecium* has more *PAK* (several hundred) K⁺ channel genes than humans (about eighty).[44] Perhaps each channel has a slightly different functional optimum. Of course, one reason for this very high number is the fact that *Paramecium* has apparently undergone several doublings in chromosome numbers in its evolutionary past. Although most of these sequences have been eliminated during subsequent evolution, many have remained, leading to a total genome about the size of the human genome.[45]

REFERENCES

1. Jennings, H. S. 1906. *Behavior of the lower organisms*. Bloomington: Indiana University Press.
2. Kung, C. 1971. Genic mutants with altered system of excitation in *Paramecium aurelia*. I. Phenotypes of the behavioral mutants. *Z. vergl. Physiol.* 71:142.
3. Kung, C. 1971. Genic mutants with altered system of excitation in *Paramecium aurelia*. II. Mutagenesis screening and genetic analysis of the mutants. *Genetics* 69:29.
4. Kung, C., et al. 1975. Genetic dissection of behavior in *Paramecium*. *Science* 188:898.
5. Naitoh, Y. and Kaneko, H. 1972. Reactivated Triton-extracted models of *Paramecium*: Modification of ciliary movement by calcium ions. *Science* 176:523.
6. Kung, C. and Naitoh, Y. 1973. Calcium-induced ciliary reversal in the extracted models of "pawn," a behavioral mutant of *Paramecium*. *Science* 179:195.
7. Satir, P. and Guerra, C. 2003. Control of ciliary motility: A unifying hypothesis. *Eur. J. Protistol.* 39:410.
8. Machemer, H. 1988. Electrophysiology. In *Paramecium*, ed. H.-D. Görtz, chap. 13. New York: Springer-Verlag.
9. Saimi, Y. and Martinas, B. 1989. Calcium dependent potassium channel in *Paramecium* studied under patch clamp. *J. Membrane Biol.* 112:79.
10. Saitow, F., Nakaoka, Y., and Oosawa, Y. 1997. A calcium-activated, large conductance and non-selective cation channel in *Paramecium* cell. *Biochim. Biophys. Acta* 1327:52.
11. Kung, C. and Saimi, Y. 1982. The physiological basis of taxes in *Paramecium*. *Annu. Rev. Physiol.* 44:519.
12. Ramanathan, R., et al. 1988. A genetic dissection of ion-channel functions. In *Paramecium*, ed. H.-D. Görtz, chap. 15, p. 236. New York: Springer-Verlag.
13. Saimi, Y. and Kung, C. 1987. Behavioral genetics of *Paramecium*. *Annu. Rev. Genet.* 21:47.
14. Haynes, W. J., et al. 1998. The cloning by complementation of the *pawn-A* gene in *Paramecium*. *Genetics* 149:947.
15. Haynes, W. J., et al. 2000. The cloning and molecular analysis of *pawn-B* in *Paramecium tetraurelia*. *Genetics* 155:1105.
16. Ogura, A. and Machemer, H. 1980. Distribution of mechanoreceptor channels in the *Paramecium* surface membrane. *J. Comp. Physiol.* 135(A):233.
17. Oertel, D., Schein, S. J., and Kung, C. 1977. Separation of membrane currents using a *Paramecium* mutant. *Nature* 268:120.
18. Saimi, Y. 1986. Calcium-dependent sodium currents in *Paramecium* mutational manipulations and effects of hyperdepolarization. *J. Membrane Biol.* 92:227.

19. Hennessey, T. M. 1987. A novel calcium current is activated by hyperpolarization of *Paramecium tetraurelia. Soc. Neurosci. Abs.* 13:108.
20. Hinrichsen, R. D., et al. 1986. Restoration by calmodulin of a Ca^{2+}-dependent K^+ current missing in a mutant of *Paramecium. Science* 232:503.
21. Kink, J. A., et al. 1991. Efficient expression of the *Paramecium* calmodulin gene in *Escherichia coli* after four TAA-to-CAA changes through a series of polymerase chain reactions. *J. Protozool.* 38:441.
22. Chan, C. W. M., Saimi, Y., and Kung, C. 1999. A new multigene family encoding calcium-dependent calmodulin-binding membrane proteins of *Paramecium tetraurelia. Gene* 231:21.
23. Saimi, Y. and Kung, C. 2002. Calmodulin as an ion channel subunit. *Annu. Rev. Physiol.* 64:289.
24. Ling, K.-Y., et al. 2001. K^+-channel transgenes reduce K^+ currents in *Paramecium*, probably by a post-translation mechanism. *Genetics* 159:987.
25. Haynes, W. J., et al. 2002. An exchanger-like protein underlies the large Mg^{2+} current in *Paramecium. Proc. Natl. Acad. Sci. U.S.A.* 99:15717.
26. Machemer, H. and Ogura, A. 1979. Ionic conductances of membranes in ciliated and deciliated *Paramecium. J. Physiol.* 296:49.
27. Satow, Y. and Kung, C. 1980. Membrane currents of *pawn* mutants of the *pwA* group in *Paramecium tetraurelia. J. Exp. Biol.* 84:57.
28. Satow, Y. and Kung, C. 1980. Ca-induced and K^+-outward current in *Paramecium tetraurelia. J. Exp. Biol.* 88:293.
29. Oertel, D., Schein, S. J., and Kung, C. 1978. A potassium channel activated by hyperpolarization in *Paramecium. J. Membrane Biol.* 43:169.
30. Kung, C. 1990. Ion channels of unicellular microbes. In *Evolution of the first nervous systems*, ed. P. A. V. Anderson, 203. New York: Plenum Press.
31. Ogura, A. and Machemer, H. 1980. Distribution of mechanoreceptor channels in the *Paramecium* surface membrane. *J. Comp. Physiol.* 135(A):233.
32. Saimi, Y. and Kung, C. 1980. Ca-induced Na^+ current in *Paramecium. J. Exp. Biol.* 88:305.
33. Saimi, Y. and Ling, K.-Y. 1990. Calmodulin activation of calcium-dependent sodium channels in excised membrane patches of *Paramecium. Science* 249:1441.
34. Preston, R. R. and Hammond, J. A. 1997. Phenotypic and genetic analysis of "chameleon," a *Paramecium* mutant with an enhanced sensitivity to magnesium. *Genetics* 146:871.
35. Machemer, H. and Teunis, P. F. M. 1996. Sensory-motor coupling and motor responses. In *Ciliates: Cells as organisms*, ed. K. Hausmann and P. C. Bradbury, 379. Stuttgart: Fischer.
36. Van Houten, J. L. 1998. Chemosensory transduction in *Paramecium. Eur. J. Protistol.* 34:301.
37. Bonini, N. M., Gustin, M. C., and Nelson, D. L. 1986. Regulation of ciliary motility in *Paramecium*: A role for cyclic AMP. *Cell Motil. Cytoskel.* 6:256.
38. Van Houten, J. L., Yang, W. Q., and Bergeron, A. 2000. Chemosensory signal transduction in *Paramecium. J. Nutr.* 130:946S.
39. Van Houten, J. L. 1977. A mutant of *Paramecium* defective in chemotaxis. *Science* 198:746.
40. Ondarza, J., et al. 2003. G-protein modulators alter the swimming behavior and calcium influx of *Paramecium tetraurelia. J. Eukaryotic Microbiol.* 50:349.
41. Paquette, C. A., et al. 2001. Glycophosphatidylinositol-anchored proteins in *Paramecium tetraurelia*: Possible role in chemoresponse. *J. Exp. Biol.* 204:2899.

42. Capdeville, Y., Cardoso De Almeida, M. L., and Deregnaucourt, C. 1987. The membrane-anchor of *Paramecium* temperature-specific surface antigens is a glycosylinositol phospholipid. *Biochem. Biophys. Res. Commun.* 147:1219.
43. Yano, J., Rachochy, V., and Van Houten, J. L. 2003. Glycosyl phosphatidylinositol-anchored proteins in chemosensory signalling: Antisense manipulation of *Paramecium tetraurelia PIG-A* gene expression. *Eukaryot. Cell* 2:1211.
44. Haynes, W. J., et al. 2003. PAK paradox: *Paramecium* appears to have more K^+ channel genes than humans. *Eukaryot. Cell* 2:737.
45. Aury, J.-M., et al. 2006. Global trends of whole genome duplications revealed by the genome sequence of the ciliate *Paramecium tetraurelia*. *Nature* 444:171.

12 Epigenetics

As originally defined in 1942, epigenetics was the genetics of development.[1] Another more recent definition of epigenetics is that it includes those mechanisms of inheritance that are not dependent on the base sequence of DNA.[2,3] This definition would exclude symbionts and mitochondria, and possibly homology-dependent inheritance. Here we will accept a more inclusive and pragmatic definition. We will define epigenesis as any inheritance whose pattern of transmission is non-Mendelian.

As might be expected, such a definition would include a wide variety of totally unrelated mechanisms, and indeed it does. The definition includes hereditary units that have their own DNA, the symbionts, and mitochondria. Cortical inheritance, a totally unrelated phenomenon that is based on the way the cortex of the cell grows and develops, is included. Serotype inheritance, completely unrelated to the other mechanisms, depends on the dynamics of protein synthesis. The remaining examples of epigenetics appear to be cases in which the phenotype is determined by the macronucleus, and the macronucleus is itself determined at the time of its formation. Only in the case of homology-dependent inheritance do we know the nature of these changes: deletion of whole regions of DNA. We know that in mating type inheritance, character differences are dependent on the differentiation of the macronuclei, but we do not know the chemical basis for these macronuclear changes. We might perhaps assume that they are due to deletions, as in the cases of homology-dependent inheritance. Indeed, the evidence is compatible with that interpretation, but we do not know whether it is true.

12.1 SYMBIONTS AND MITOCHONDRIA

Symbionts and mitochondria represent a very large class of cases of non-Mendelian inheritance in *Paramecium*. All show the cytoplasmic pattern of transmission. See Chapter 6. Not only do the symbionts and mitochondria have their own DNA, but many also harbor a second layer of infection in the form of incomplete phages and plasmids within the bacteria-like forms. Many kappas also have a small structure called an R body, which is taken in by mouth and suddenly unrolls in the food vacuoles during killing. R bodies appear to be necessary for the toxic action of killers on sensitives, but exactly how is not known. An unexplained feature of the killer phenomenon is how the presence of the symbiont within the killer makes it resistant to the toxic effect of the killer. Another unknown is the chemistry of the maintenance genes for each symbiont.

Apparently *Paramecium* has been invaded numerous times in its evolutionary history by many different symbionts. Most symbionts are obligate, while others can be cultured outside of *Paramecium*. These many cases provide one of the prominent kinds of epigenetic inheritance in *Paramecium*.

12.2 SEROTYPE INHERITANCE

The serotype system of *Paramecium* consists of a series of related proteins on the surface of the cell that show mutual exclusion, that is, only one serotype gene can be expressed at a time. See Chapter 7. The genes for these proteins form a family of a dozen or more related genes in most stocks. The proteins are very large, very numerous, and strongly antigenic. For this reason, the injection of only a few paramecia into a rabbit will often produce only one major antibody, directed against the major protein expressed. Exposure of paramecia to homologous antiserum results in immobilization of the cells. The presence of one of the proteins on the cell surface results in a distinct antigenic type or serotype. Serotypes can switch from one to another. Although each has an optimum condition for expression (temperature, growth rate, or other environmental conditions), usually a single condition can be found that allows most of the serotypes to be maintained. Cytoplasmic inheritance is seen after crosses between the serotypes. The situation seems to devolve into the control of gene action. Although it has been possible to determine sensitive regions of the cloned genes, no answers to the nature of the control have been found. The phenomenon seems comparable to the situation in the development of multicellular organisms whose genes may be turned off and on, in a rather stable way. The answers to serotype inheritance seem to be hidden in the larger puzzle of the control of gene action.

12.3 MATING TYPE INHERITANCE

Mating type in *Paramecium* is inherited in different ways in different strains. See Chapter 5. In one group (C) it is determined by simple Mendelian inheritance, and it therefore is not epigenetic. In the second group (A) its pattern of transmission is caryonidal. See below for explanation of caryonidal. In the last group (B) its pattern of inheritance is still weakly caryonidal, but is strongly influenced by the cytoplasm, and transmission is primarily cytoplasmic. In spite of this rather major difference in the patterns of inheritance between groups A and B, the two kinds of inheritance are both thought to be closely related and depend on the macronucleus. The nature of these nuclear differentiations is unknown.

12.3.1 Caryonidal Inheritance of Mating Type

Although caryonidal inheritance represents the only type of epigenetic inheritance that does not follow the cytoplasm, it clearly does not follow the simple rules of Mendelism either. It is free of all cytoplasmic influences. The primary character inherited in this way is mating type, although Sonneborn believed that other cases followed this same pattern. An example was found in the very early experiments of Jollos (see Sonneborn,[4] p. 308). During autogamy and conjugation, two new macronuclei are formed and then segregated from each other at the first fission after autogamy or conjugation. Each of the two new cells and all of its progeny is called a caryonide. Thus, each of the two nuclei forms a new caryonide, and each is independently determined with a certain probability to be one of two of the alternative

Epigenetics

mating types. Each caryonide has a mating type that breeds true to type until its macronucleus is lost at the next autogamy or conjugation.

12.3.2 CYTOPLASMIC INHERITANCE

In group B at conjugation or autogamy one sees cytoplasmic transmission of mating type. Thus, the cytoplasm influences the determination of the two new macronuclei that are formed in each cell to cause each macronucleus to become like its parent. It has also been shown that while cytoplasmic inheritance generally determines the mating type, it is possible for a line of cells to acquire an allele that will overide the cytoplasmic effect permanently and produce an alternative mating type. When this happens, it has been found that removal of the gene at any subsequent time by a cross causes the line of cells to return to its original mating type. See, for example, the work of Taub and also that of Brygoo described in Chapter 5.

12.4 CORTICAL INHERITANCE

The cortex of *Paramecium* appears to be self-perpetuating. See Chapter 10. Cortical inheritance is clearly found in numerous ciliates, and the principles on which it is based may well be found in various systems in other organisms. See Grimes and Aufderheide[5] and Frankel.[6] It occurs in many, if not all, ciliated protozoa because the cortex consists of cells of old asymmetrical structures that form the pattern for the development of new structures.

12.5 HOMOLOGY-DEPENDENT INHERITANCE

The presence of all these different types of epigenetic phenomena found in *Paramecium* (symbiosis, gene expression, the inheritance of mating type, and cortical inheritance), and in greater numbers than are present in any other organism, certainly justifies Sonneborn's view of the major importance of the cytoplasm in inheritance in *Paramecium*. But the support for this view is actually much stronger than indicated by these cases of epigenetics. As shown below, the cytoplasmic inheritance exemplified by the d48 mutant is by itself enough to justify Sonneborn's beliefs, for it can affect any gene-controlled character not necessary for life.

12.5.1 D48: A HOMOLOGY-DEPENDENT MUTANT

The story began when Epstein and Forney[7] started a search for mutants of the A^{51} gene in stock 51 of *Paramecium tetraurelia*. They screened for mutations unable to express serotype A. They found several mutations that reduced or eliminated serotype A. Most of them showed Mendelian inheritance, either at the A^{51} locus or in other genes. An example is the mutant d12, which proved to be a simple deletion of the *A* gene with normal Mendelian inheritance. The gene was deleted in both micronucleus and macronucleus. By this time, techniques had been perfected for the microinjection of fluids into *Paramecium* and the transport of nuclei between cells. It was found that the microinjection of the cloned wild-type A^{51} gene with its flanking regions into the macronucleus of d12 cured the mutant temporarily and allowed the expression of the A^{51} gene.[8] Thus, injected circles of DNA were linearized, acquired telomeres, and reproduced, and were expressed in *Paramecium*. However, after the next autogamy the old injected macronucleus bearing the injected A^{51} gene was lost and the phenotype reverted to mutant.

Another mutant was called d48. It had the A^{51} gene in its micronucleus, but remarkably, the gene was deleted from the macronucleus. Moreover, d48 was inherited cytoplasmically. When the same experiment was tried with d48, that is, when the A^{51} gene was injected into the d48 mutant, again a cure in the deficiency of gene A occurred and serotype A was expressed, but this time the cure did not end with the next autogamy, but was permanent.[9,10] Even a small section of the gene was capable of producing this rescue.

This finding led to the hypothesis that the only defect in d48 is the absence of the wild-type gene for serotype A in the macronucleus, and that the presence of the gene in the old macronucleus ensures its presence in the newly forming macronucleus. This hypothesis was confirmed by the replacement of the micronucleus in d12 (remember that d12 is a Mendelian deficiency in the A gene in both the micronucleus and macronucleus) with a micronucleus from wild type, thereby creating a new mutant like d48 with wild-type micronuclei and a deficient macronucleus.[11] This same experiment was repeated by Scott et al.[12] with the B gene by transferring a wild-type nucleus into a Mendelian mutant strain lacking the B gene, and producing a d48-like strain deficient in the B gene. The A gene remained wild type during this manipulation. Further work showed that the A-deficient d48 and the B-deficient strains were independently inherited.[13,14]

Attempts have been made to determine the smallest region of the A gene that can induce a restoration of d48 to wild type. It is found that there is a section consisting of several thousand bases near the start of the gene that is effective. The most effective part of this region is near its middle. It includes a series of repeats, approximately one thousand bases in length. The larger the region injected, the more effective it is.[15] The effects of injecting two separate regions are greater than injecting only one of the two.

It is also noted that another complexity besides simple cytoplasmic inheritance is evident in the crosses. Evidence is provided by Rudman et al.[16] that the micronucleus is also involved in the cytoplasmic effect. This evidence consists of crossing d12 × d48 and d12 × wild type and comparing the F1 phenotypes of the progeny of the d-12 exconjugants. According to the theory just given, since d48 and wild type both have wild-type micronuclei, the two micronuclei should produce the same effect in d12. Instead, the two F1s produced by the two crosses are not alike in the F2 progeny of the crosses in the d12 exconjugant, for the micronucleus from d48 is much more likely to produce deletions than the micronucleus from wild type. This result shows that the capacity to produce a cytoplasmic effect is found not only in the macronuclei and cytoplasm of d48, but also in the micronucleoplasm of d48. One recalls that a similar effect was shown for micronuclei on mating type transmission in Chapter 5. The PCR evidence provided in this same paper that d12 has the A^{51} gene in its micronucleus was found to be incorrect. Direct blots of the DNA from isolated d12 micronuclei provided incontrovertible evidence that d12 lacks the A^{51} gene.[17] Presumably, the erroneous PCR-based data that led to the faulty conclusion occurred because of contamination, either by DNA in the PCR test or by the cultures used for the DNA.

We might speculate on the mechanism by which homology-dependent inheritance operates. One concludes that in autogamy in d48 some factor must move from the old macronucleus to the newly forming macronucleus and make the genes in the new macronucleus like those in the old macronucleus. Moreover, the same process presumably occurs in wild type. The specificity of this process suggests that the substance is most likely nucleic acid, hence the name homology-dependent inheritance. However, the precise mechanism by which homology-dependent inheritance works is unknown.

12.5.2 Induction and Repair of Homology-Dependent Mutants

Paramecium is unusual in that telomeres do not always form at the same point in the chromosomes of all wild-type individuals. Telomere addition regions (TARs) in wild type are small chromosomal regions consisting of a number of bases where chromosomes may end and telomeres form. Each TAR is a few hundred bases long. Moreover, there are four alternate TARs in *P. tetraurelia* near the end of the chromosome where the *A* gene resides spread over a region of 2 to 10 kb.[18] Thus, at macronuclear formation, a TAR is selected first and then the precise position of the telomere within the TAR is chosen.[19,20] The breaks in the chromosome seen in the mutants can then be understood in terms of TARs, for the deletions are seen as breaks at new TARs, upstream of the wild-type ones. This holds for the gene-controlled as well as the d48 deletions. The position of the break within a TAR is never precise.

Posttranscriptional gene silencing (PTGS) occurs in *Paramecium* and has been studied in many organisms.[21,22] PTGS is seen in *Paramecium* when DNA is injected into the macronucleus of vegetative wild-type cells. Examination before autogamy reveals that the corresponding resident genes of the macronucleus are suppressed. This is a remarkable phenomenon and appears to hold for virtually all the genes in the genome. The gene remains silent for as long as the injected gene is present in the *Paramecium* and is then lost at the next autogamy or conjugation. Although it may persist for a long period until autogamy occurs, it persists in this period because the inducing DNA is continuously replicated and transcribed. It has also been found that the introduction of double-stranded RNA homologous to the gene will also induce PTGS,[23] showing clearly that induction occurs at the level of RNA. No injections are required, for the cells simply take up the RNA by feeding. This procedure provides an easy way of inducing PTGS in *Paramecium*.

While PTGS is a temporary phenomenon, the induction of mutations that appear after autogamy is permanent. It has been found that whenever PTGS is observed, at the next autogamy permanent cytoplasmic mutants are induced. See below. The two phenomena are perfectly correlated. Whenever mutants are produced after autogamy, the mutants are always preceded by PTGS of the relevant genes.[20] This relation still holds when double-stranded RNA is used to produce PTGS, for cytoplasmic mutants can be found after autogamy in this case as well.[20]

These experiments on d48 constitute a new way of producing epigenetic mutants. The original observations were made by Meyer.[24] Here it was found that injection of various sequences into the macronucleus of wild-type *P. primaurelia* induced their deletion. At the next and subsequent autogamies after injection, although deletions in the same section of DNA that was injected appeared, the TARS varied in position. However, with successive autogamies they eventually became more stable—different stable lines having different TARS, each representing a different mutant. Moreover, these deletions were inherited cytoplasmically exactly like d48.[25] These observations were extended in other papers; see Le Mouël et al.[26] Injection of all sequences of DNA that were tried resulted in elimination of those specific sequences after autogamy.

Later it was discovered by Garnier et al.[20] that it made a difference whether the injected gene was expressed during the first autogamy after it was injected. Expression of the injected gene during the first autogamy after injection repressed PTGS and mutant production, while nonexpression had the opposite effect. Thus, A^{29}, which is not often expressed, when injected into stock 29, readily induced PTGS and mutants, while A^{51}, which is readily expressed, when injected into stock 51, did not. The notion that expression or nonexpression was the determining factor was borne out by two additional observations. First, in the case of the experiment with A^{29} that induced PTGS and mutants in most lines, there were two lines in one experiment where no silencing of mutants occurred. Inspection revealed that these were two lines in which the injected gene was expressed. Second, they injected a modified form of the A^{51} gene that could not be expressed because it contained a reading frame mutation. It readily induced PTGS and mutants when injected into stock 51.

In a study of PTGS, Galvani and Sperling[22] found that injection of a whole gene with both 5′ and 3′ flanking regions did not induce silencing. However, if they omitted the 3′ flanking regions, then PTGS was induced. This finding was true for all the different genes tried.

There is still a contradiction that needs explanation, for now we see that injection of a nonexpressible A^{51} gene into a wild-type cell creates an A^{51} mutant at the next autogamy after injection, while injection of the same DNA into a mutant produces wild type. Epistatic mutants like d48 can either be *produced* when A^{51} is injected into wild type or *rescued* when A^{51} is injected into a mutant. In an experiment designed to examine the process of rescue, the A^{29} gene was injected into an A^{29} mutant at different levels of multiplicity. At low multiplicities curing was obtained (rescue, TARs shifted downward). However, at high multiplicities more extreme deletions were produced (mutations induced, TARs shifted upward), so that in one experiment both deletions and curing could be seen.[20]

The curing of deletions and the induction of new mutants are both highly specific. Thus, when the A^{51} gene was injected into serotype B of stock 51, subsequent deletions in A^{51} were very rare, but occasionally were induced. When *B* was injected into stock 51, B^{51} deletions were induced frequently. So induction of A and B deletions is specific. The same was true of the rescue of *B*. Moreover, the two mutants were inherited independently. So mutation induction and rescue are both specific.[12] These facts suggest that the deletions and rescues are dependent on a highly specific mechanism, most likely the base sequence of sections of the DNA being manipulated.

As already noted, experiments carried out using this technique, and microinjection as well, showed that after the first induction of a mutant, several autogamies may be necessary before a stable mutant can be selected. This ability of TARs to move may account for some of the different types of TARs obtained in the Mendelian and non-Mendelian mutants investigated by Rudman et al.[16]

12.5.3 Homology-Dependent Mutants That Fail to Excise IESs

Mutants of this type show epigenetic inheritance. In wild-type stock 51 of *P. tetraurelia* genotype *mtF*[+], the serotype G gene contains an IES that, as expected, is eliminated like all IESs at autogamy; the phenotype of this strain is said to be IES[−], since it has the IES missing. On the other hand, a Mendelian mutant of *P. tetraurelia*, *mtF*[E], retains this IES and is therefore designated phenotype IES[+].[27] See Chapter 5. The *mtF*[E] gene is pleiotropic as well and affects numerous other cellular activities, including viability and mating type. It was isolated as a mutant that was pure for mating type E, hence the name. A cross of *mtF*[E]/*mtF*[E] (mutant, mating type E) with *mtF*[+]/*mtF*[+] (wild type, mating type O) after autogamy from each of the two exconjugants from a pair produces, as expected, typical Mendelian inheritance: half are *mtF*[E]/*mtF*[E] and half are *mtF*[+]/*mtF*[+]. However, both mating type and the properties of the mutant are cytoplasmically inherited. Thus, all the individuals from the mating type E side of the cross after autogamy retained the properties of the mutant, inherited cytoplasmically from the mutant member of the original cross. These IES[+] segregants from the E side of the cross retain their original phenotype and, although having the gene *mtF*[+]/*mtF*[+], are IES[+]. Crosses of the cytoplasmic mutant (*mtF*[+]/*mtF*[+], IES[+]) to wild type (*mtF*[+]/*mtF*[+], IES[−]) show cytoplasmic inheritance.

Yao and his group found similar mutants in *Tetrahymena*. The M and R elements are present in wild type as IESs that are normally eliminated at conjugation in *Tetrahymena*. Wild type is designated IES M[−] and IES R[−]. However, when macronuclei of IES M[−] and IES R[−] *Tetrahymena* are injected with large doses of M or R DNA, epigenetic mutants affecting DNA processing, IES M[+] and IES R[+], are produced.[28–31] Theories to explain the results in *Tetrahymena* are given in Yao and Chao.[32]

12.5.4 The Elimination of Foreign DNA

One also notes that if a foreign DNA is injected into the micronucleus of *Tetrahymena*, it is eliminated at the next conjugation. Thus, a mechanism is provided to protect the cell from invading foreign DNA.[33] Yao and Chao[32] give theories on the mechanism of such deletions.

12.5.5 Theories of PTGS

Galvani and Sperling[22] and Yao and Chao[32] have presented models for how PTGS works in *Paramecium*. The process is very complex and many elements are involved. It is now generally conceded that RNAi (RNA interfering) sequences are the basis for the deletions. Some of the elements that must be considered in the silencing process are as follows. First, the injected DNA produces double-stranded RNA.[22] This RNA is broken into small pieces by the ribonuclease enzyme called Dicer. These short pieces of double-stranded RNA may produce single-stranded RNA. Small (~22–23 nt) RNA fragments accumulate in *Paramecium* during PTGS.[20] Madireddi et al.[34] found that pieces of RNA are joined by proteins called Argonaute proteins to form the RNA-induced silencing complex (RISC). The programmed DNA degradation (PDD) proteins are also involved. PDD1 was formerly known as protein 65.[34,35] It is associated with heterochromatin and gene silencing. Histone H3 lysine 9 also must be methylated to produce elimination in *Tetrahymena*.[36]

It is difficult to account for all these multiple phenomena in one theory. However, a simple view is that the injected DNA produces large quantities of transcripts that are unlike the normal gene mRNA, but they still can hybridize with it and neutralize the activity of the mRNA, and thereby silence it. See numerous references in Galvani and Sperling.[22]

12.6 CONCLUSIONS

It is evident that any mutation that can be induced can occur spontaneously as a rare event. In that case, any gene that is not necessary for the life of the cell should be able to undergo a d48-like mutation. Isolation of the gene and molecular work are necessary to be sure that we are dealing with such cases. However, whenever a cytoplasmically inherited character in a trait not essential for life appears, it is suspect. One such example is that of a behavioral mutant showing caryonidal inheritance.[37] Another example of a cytoplasmic mutant is the trichocyst mutant, d4-113, studied by Sonneborn and Schneller.[38] Many other cases from the older literature meet this requirement, and homology-dependent mutants must be considered one of the major epistatic phenomena found in *Paramecium*. Mutants of this type are likely to be found in any nonessential gene.

We are now in a position to point out the attributes of homology-dependent inheritance, a truly remarkable kind of epigenetic inheritance that can occur in almost any trait in *Paramecium*. (1) The alternative traits are determined by an effect of the old macronucleus on the newly forming macronucleus at autogamy or conjugation. The traits are manifested by the action of the macronucleus, not the micronuclei. (2) These traits are transmitted by an apparent cytoplasmic inheritance. It has also been found that the substance transmitted through the cytoplasm and originating in the old macronucleus is nucleic acid, either RNA or DNA, which can protect the newly forming nucleus, yielding a new role for nucleic acid in development. (3) The alternative traits are produced by an elimination or retention of specific sequences of DNA from the new macronucleus when it is formed from the micronucleus, with nucleic acid from the old macronucleus determining whether sections of DNA are cut out or protected. We have cited numerous examples of homology-dependent inheritance in *Paramecium*, and have also seen numerous other cases that have puzzled investigators in the past that may well be additional examples.

It is interesting to note the derivation of the phrase "homology dependent," implying the need for nucleic acid base pairing. It appears to have been applied by investigators first in their description of PTGS in plants. The term was used to describe the cases of stable hereditary changes in *Paramecium* by Meyer and Duharcourt,[39] and now appears to be gaining acceptance as the phrase of choice to describe such phenomena in ciliates.[40]

Considering the numerous kinds of epigenetic inheritance exhibited by *Paramecium*, undoubtedly more than in any other organism, it is no wonder that ciliate geneticists have been confused by their peculiar little organisms. However, we think now that we are beginning to perceive the true picture of what goes on in *Paramecium* with its micronuclei and macronuclei. The work on ciliates was not considered in early accounts of PTGS and epigenetics.[41] However, with all the current work on ciliates, the situation has changed. Perhaps this work will lead to a better understanding of the phenomenon in higher organisms, and even help solve the problem of differentiation of cells in development. This was the problem posed by Waddington in his original definition of epigenesis many years ago.[1]

REFERENCES

1. Waddington, C. H. 1942. The epigenotype. *Endeavor* 1:18.
2. Lederberg, J. 1958. Genetic approaches to somatic cell variation: Summary comment. *J. Cell. Comp. Physiol.* 52(Suppl. 1):383.
3. Lederberg, J. 2001. The meaning of epigenetics. *The Scientist* 15:6.
4. Sonneborn, T. M. 1947. Recent advances in the genetics of *Paramecium* and *Euplotes*. *Adv. Genet.* 1:263.
5. Grimes, G. W. and Aufderheide, K. J. 1991. *Cellular aspects of pattern formation: The problem of assembly.* Basel: Karger.
6. Frankel, J. 1989. *Pattern formation: Ciliate studies and models.* New York: Oxford University Press.
7. Epstein, L. M. and Forney, J. D. 1984. Mendelian and non-Mendelian mutations affecting surface antigen expression in *Paramecium tetraurelia*. *Mol. Cell. Biol.* 4:1583.
8. Godiska, R., et al. 1987. Transformation of *Paramecium* by microinjection of a cloned serotype gene. *Proc. Natl. Acad. Sci. U.S.A.* 84:7590.
9. Harumoto, T. 1986. Induced change in a non-Mendelian determinant by transplantation of macronucleoplasm in *Paramecium tetraurelia*. *Mol. Cell. Biol.* 6:3498.
10. Koizumi, S. and Kobayashi, S. 1989. Microinjection of plasmid DNA encoding the A surface antigen of *Paramecium tetraurelia* restores the ability to regenerate a wild type macronucleus. *Mol. Cell. Biol.* 9:4398.
11. Kobayashi, S. and Koizumi, S. 1990. Characterization of non-Mendelian and Mendelian mutant strains by micronuclear transplantation in *Paramecium tetraurelia*. *J. Protozool.* 37:489.
12. Scott, J., Leeck, C., and Forney, J. D. 1994. Analysis of the micronuclear B type surface protein gene in *Paramecium tetraurelia*. *Nucl. Acids Res.* 22:5079.
13. Scott, J. M., et al. 1994. Non-Mendelian inheritance of macronuclear mutations is gene-specific in *Paramecium tetraurelia*. *Mol. Cell. Biol.* 14:2479.
14. You, Y., Scott, J., and Forney, J. D. 1994. The role of macronuclear DNA sequences in the permanent rescue of a non-Mendelian mutation in *Paramecium tetraurelia*. *Genetics* 136:1319.
15. Kim, C. S., Preer, J. R. Jr., and Polisky, B. 1994. Identification of DNA segments capable of rescuing a non-Mendelian mutant in *Paramecium*. *Genetics* 136:1325.
16. Rudman, B., et al. 1991. Mutants affecting processing of DNA in macronuclear development in *Paramecium*. *Genetics* 129:47.
17. Preer, L. B., Hamilton, G., and Preer, J. R. Jr. 1992. Micronuclear DNA from *Paramecium tetraurelia*: 51A gene has internally eliminated sequences. *J. Protozool.* 39:678.
18. Amar, L. and Dubrana, K. 2004. Epigenetic control of chromosome breakage at the 5′ end of *P. aurelia* gene A. *Eukaryot. Cell.* 3:1136.

19. Forney, J. D. and Blackburn, E. H. 1988. Developmentally controlled telomere addition in wild type and mutant paramecia. *Mol. Cell. Biol.* 8:251.
20. Garnier, O., et al. 2004. RNA-mediated programming of developmental genome rearrangements in *Paramecium*. *Mol. Cell. Biol.* 24:7370.
21. Ruiz, F., et al. 1998. Homology-dependent gene silencing in *Paramecium*. *Mol. Biol. Cell* 9:931.
22. Galvani, A. and Sperling, L. 2001. Transgene-mediated post-transcriptional gene silencing is inhibited by 3′ non-coding sequences in *Paramecium*. *Nucl. Acids Res.* 29:4387.
23. Galvani, A. and Sperling, L. 2002. RNA interference by feeding in *Paramecium*. *Trends Genet.* 18:11.
24. Meyer, E. 1992. Induction of specific macronuclear developmental mutations by microinjection of a cloned telomeric gene in *Paramecium primaurelia*. *Genes Dev.* 6:211.
25. Meyer, E., et al. 1997. Sequence-specific epigenetic effects of the maternal somatic genome on developmental rearrangements of the zygotic genome in *Paramecium primaurelia*. *Mol. Cell. Biol.* 17:3589.
26. Le Mouël, A., et al. 2003. Developmentally regulated chromosome fragmentation linked to imprecise elimination of repeated sequences in paramecia. *Eukaryot. Cell* 2:1076.
27. Duharcourt, S., Butler, A., and Meyer, E. 1995. Epigenetic self-regulation of developmental excision of an internal eliminated sequence in *Paramecium tetraurelia*. *Genes Dev.* 9:2065.
28. Chalker, D. L. and Yao, M. C. 1996. Non-Mendelian, heritable blocks to DNA rearrangement are induced by loading the somatic nucleus of *Tetrahymena thermophila* with germ line-limited DNA. *Mol. Cell. Biol.* 16:3658.
29. Chalker, D. L. and Yao, M. C. 2001. Nongenic bidirectional transcription precedes and may promote developmental DNA deletion in *Tetrahymena thermophila*. *Genes Dev.* 15:1287.
30. Yao, M. C., Fuller, P., and Xi, X. 2003. Programed DNA deletion as an RNA-guided system of genome defense. *Science* 300:1581.
31. Liu, Y., Mochizuki, K., and Gorovsky M. A. 2004. Histone H3 lysine 9 methylation is required for DNA elimination in developing macronuclei in *Tetrahymena*. *Proc. Natl. Acad. Sci. U.S.A.* 101:1679.
32. Yao, M.-C. and Chao, J.-L. 2005. RNA-guided DNA deletion in *Tetrahymena*: An RNAi-based mechanism for programed genome rearrangements. *Annu. Rev. Genet.* 39:537.
33. Liu, Y., et al. 2005. Elimination of foreign DNA during somatic differentiation in *Tetrahymena thermophila* shows position effect and is dosage dependent. *Eukaryot. Cell* 4:421.
34. Madireddi, M. T., et al. 1996. Pdd1p, a novel chromodomain-containing protein, links heterochromatin assembly and DNA elimination in *Tetrahymena*. *Cell* 87:75.
35. Jenuwein, T. and Allis, C. D. 2001. Translating the histone code. *Science* 293:1074.
36. Liu, Y., Mochizuki, K., and Gorovsky, M. A. 2004. Histone H3 lysine 9 methylation is required for DNA elimination in developing macronuclei in *Tetrahymena*. *Proc. Natl. Acad. Sci. U.S.A.* 101:1679.
37. Rudman, B. M. and Preer, J. R. Jr. 1996. Non-Mendelian inheritance of revertants of paranoiac in *Paramecium*. *Eur. J. Protistol.* 32:141.
38. Sonneborn, T. M. and Schneller, M. V. 1979. A genetic system for alternative stable characteristics in genomically identical homozygous clones. *Dev. Genet.* 1:21.
39. Meyer, E. and Duharcourt, S. 1996. Epigenetic programming of developmental genome rearrangements in ciliates. *Cell* 87:9.
40. Preer, J. R. Jr. 2006. Sonneborn and the cytoplasm. *Genetics* 172:1373.
41. Russo, V. E. A., Martienssen, R. A., and Riggs, A. D., eds. 1996. *Epigenetic mechanisms of gene regulation*. Monograph 32. Cold Spring Harbor, NY: Cold Spring Harbor Laboratory Press.

Index

A

Algae, 6, 21, 24, 64, 68
Amacronucleate cells, 36, 139
American Type Culture Collection, 7, 21
Amicronucleate, 38
Amicronucleate cells, 139
Amino acids, 117, 119, 122, 123
    of immobilization antigens, 116
    periodicity, 120, 122
Amitosis, 30, 58
Amoebozoa, 19, 20
Antibiotics, 70, 88, 89
Antibodies, 100, 101; *see also* Immobilization antigens
Antigenic type, 47
Antigenic variation, 6, 99, 106, 107, 126
Antigens, *see* Immobilization antigens
Archaea, 20
Archaeplastida, 20
Archibacteria, 20
Asexual fission, 29–30, 65
Autogamy, 11, 34–35
    and caryonidal transmission, 45, 52
    and gene silencing, 47
    genetic effects, 35
    induction, 35, 38
    in old age, 38
    in *P. caudatum*, 35
    and rejuvenation, 37
    in selfers, 55
Avoiding reaction, 167–168

B

Bacteria, 20, 76; *see also* Symbionts
    for culturing paramecia, 43
Basal body, 22, 27
Basal granule, 22, 158
Bateson, W., 1
Behavior
    avoiding reaction, 167–168
    folate response, 170
    mutants, 46, 165
    response to stimuli, 169–170
    reverse swimming, 166
    touch response, 167, 168, 170
Behavior of the Lower Organisms, 165

C

Ca^{2+} channels, 168
    function of calmodulin, 169
Caedibacter caryophila, 82
Caedibacter paraconjugatus, 77
Caedibacter taeniospiralis, 68, 73
Caedibacter varicaedens, 70, 71, 80
Caedobacter taeniospiralis, 68, 78
Calmodulin, 168, 169
    CAM gene coding, 49
    curing mutants, 170–171
Caryonidal inheritance, 33, 45–46
    versus cytoplasmic, 46
    mating types, 51–53, 176–177
Caryonides, 33, 34
    defined, 45
Cell membranes
    electric properties, 166–167
Centrins, 160
Cha (chameleon) mutant, 169
Chemotaxis, 165
Chloroplasts, 6, 64
Chromalveolata, 20
Cilia, 21
    cross section, 25, 26
    development, 22
    and immobilization, 101
    and immobilization-antigens, 127
    in inverted kineties, 158
    microtubules, 26
    sliding filament model, 166
Ciliates
    genic balance, 30
    life cycle, 36–38
    nomenclature, 17
    nuclear dualism, 145–146
    nuclei, 3
    predators, 23–24
    structural organization, 3
Circadian clocks, 59–60
Clones
    combinations, 11, 12
    preparation, 43–44
Cloning, 7
Cnr (ciliary nonreversible) mutants, 168
Coenorhabditis elegans, 2
Colpidium, 157

Conjugation, 11
  and caryonidal transmission, 45, 51, 52
  discovery, 12
  doublet formation, 32
  duration, 32
  environmental conditions, 32
  exchange of cytoplasm, 34
  exchange of nuclei, 33
  in life cycle, 36
  nuclear changes, 31
  process of, 30, 32–34
  and rejuvenation, 30, 37–38
  role of paroral cone, 24
Cortex, 21–24
Cortical inheritance, 6, 47, 160, 177
Cortical mutants, 158–160
C.pseudomutants, 78
Crossbreeding, 4
Culture mediums, 43
Cytogamy, 36, 41, 59
Cytoplasmic transmission, 6, 34, 46–48
  d48 mutation, 47
  early views of, 63, 64
  of Kappa, 65
  versus macronuclear, 46
  and mating type, 53–54, 177
  and selfers, 54–56
  and serotype variants, 47
Cytotaxis, 160

D

Dancer mutation, 168
Darwin, 1
Didinium, 23, 84, 102
Dileptus, 23
D48 mutant, 47, 115, 177–179
DNA, 1, 2; *see also* Ribosomal DNA (rDNA)
  deletions, 181
  foreign, 181
  injection into macronuclei, 7, 49
  methylation, 6
Double animals, 157–158
Drepanospira Mülleri, 81
Drosophilia, 1, 3
  cellular generations, 2
  and conjugation, 30
  CO_2-sensitive, 66
  heredity, 63–64
  life cycle, 4

E

E. coli, see *Escherichia coli*
Embryogenesis, 6
Endomixis, 11, 34

Epigenetics, 6, 182–183; *see also* Cytoplasmic transmission
  definitions, 175
  and mitochondria, 175
  states, 115
  and symbionts, 175
  types of phenomena, 175
Erythromycin resistance, 88, 89
Escherichia coli, 3, 48, 70, 118, 119
  calmodulin expression, 169
  life cycle, 4
  plasmids, 145
  R bodies, 75
Eubacteria, 20
Eucaryotes, 20
Eukaryotes, 2
  chromosome number, 145
  classification, 20
Euplotes, 17
Evolution, 3, 19, 155–156, 171
Excavata, 20

F

Findly and Gall
  research by, 151–152, 154
Fission, 29–30
  first stage, 46
  and gene expression, 45
  and life cycle periods, 36–38
  and variation, 2
Flatworms, 157
Functional complementation, 7, 49, 168

G

Galavanotaxis, 165
Gall, J.G., research by, 151–152, 154
Gene cloning, *see* Cloning
Gene identification
  based on protein sequence, 49
  based on protein similarity, 49
  functional complementation, 49
  MALDI-TOF analysis, 49–50
  using RNAs, 48
Gene silencing, 7, 47, 48, 179–181
Genetics of Paramecium aurelia, 5
Genetics of the Protozoa, 3
Genetics Society of America, 5
Genomes
  immobilization antigens, 119
  of P. tetraurelia strain d4-2, 48, 145, 171
  size in *Paramecium,* 171
Geotaxis, 165
Germ cells, 1, 67
Guanine monophosphate (GMP), 161

Index

H

Hemixis, 36
Heterocaryons, 139–140
Heterocaryon test, 46
Heterokaryons, 36
Holospora caryophila, 79–81
Holospora elegans, 81
Holospora obtusa, 81
Holospora undulata, 81
Homology-dependent inheritance, 7, 48, 175, 177
    and D48 mutant, 177–179
    and mating type inheritance, 60
Human Genome Project, 2
Hypotrichs, 142

I

I-antigens, *see* Immobilization antigens
IESs, *see* Internal eliminated sequences (IESs)
Immobilization antigens, 99–100
    abundance on surface, 101
    allelic exclusion, 115
    allelic variations, 110, 112–113, 122–123
    amino acid compositions, 116
    and chemoresponses, 102
    conjugation between serotypes, 105
    control by cytoplasm, 104
    control of cytoplasm, 113–114
    environmental control, 104, 106–110
    functions in cell, 102
    gene expression, 124, 126
    gene promoters, 129, 131
    gene sequencing, 117–119, 120–124
    gene structure, 128
    genetic code, 119
    genetic features, 102–103
    heterozygotes and dominance, 115
    immobilization test, 100–102
    isogenes, 124
    isolation, 115–117
    measuring concentration, 102
    mRNAs, 117
    mutual exclusion test, 131–132
    nomenclature, 102, 103
    nonallelic variations, 122–123
    positional control, 126–127
    run-on transcription, 128–129
    serotypes, 102, 103
    serotype switching, 127–128
    variability, 102
Immunization, 100
Incompatibility groups, 13; *see also* Mating types
Internal eliminated sequences (IESs), 3, 141–142
    excision models, 142
    and homology-dependent mutants, 181
    in hypotrichs, 142
    mutations within, 142
    removal and mating type, 60
Ion channels
    Ca^{2+}, 167, 168
    cell membrane properties, 166–167
    identifying channel proteins, 170
    K^+, 168–169
    Mg^{2+}, 169
    Na^+, 169
    patch clamp studies, 167, 170–171
    role of mutants, 168

J

Jennings, H., 3

K

Kappa, 6, 33, 64; *see also* Killer paramecia
    classification, 67
    cytoplasmic exchange, 34, 66
    and gene *K,* 66, 67, 71, 82–83
    genome, 86
    growth, 67
    infection of, 74
    killing properties, 72, 80, 175
    nomenclature, 68
    plasmagene hypothesis, 64–68
    with R bodies, 71
    reproduction, 72–73
    structure, 74
    variation, 76
K^+ channels, 168–169
Killer paramecia, 64–68
    cross with sensitives, 65, 66, 71
    hybridization studies, 71
    mate killers, 76–77
    in *P. biaurelia,* 70–71
    in *P. tetraurelia,* 73–76
    paralysis, 78
    rapid lysis, 77
    resistance, 72
    spinner, 78
Kinetodesmal fibers, 22, 23, 27, 158
Kinetosomes, 22, 25, 27, 158, 160
    development stages, 28
Kinety, 22, 23
    inverted mutant, 159
Klebsiella aerogenes, 43

L

Life cycles, 4
    cause of senescence, 38
    immaturity, 36

maturity, 37
old age, 37
rejuvenation, 37–38
Lyticum flagellatum, 77
Lyticum sinuosum, 77

M

Macronuclear inheritance, *see* Caryonidal inheritance
Macronuclear regeneration (MR), 35–36, 44
   and mating type, 53
Macronuclei, 24–25, 26
   activity of, 139–140
   anlagen, 33, 34
   chromosomal breaks, 140–141
   disintegration, 32, 33
   division, 30
   DNA injection, 7
   formation, 11, 26, 33, 35–36
   genic balance at division, 143–145
   irregularities, 38
   versus meganuclei, 25
   and phenotype, 11, 30, 36
   ploidy levels, 139, 146
   rDNA, 152–153
   size, 26
   telomeres, 140–141
MALDI-TOF analysis, 49–50
Mate killers, 76–77
Mating types, 4, 5
   and caryonidal inheritance, 51–53, 176–177
   complementary, 44
   cross-reactions, 12, 13, 14, 51
   and cytoplasmic state, 54–55
   and cytoplasmic transmission, 53–54, 177
   discovery, 4, 11–12
   homologies, 51
   homology-dependent inheritance, 60
   micronuclear determination, 57
   nomenclature, 13, 16
   P. bursaria, 58
   P. caudatum, 58–59
   P. multimicronucleatum, 59–60
   P. primaurelia, 52–53
   and serotype, 57
   by species, 12, 51
   switching phenomenon, 59–60
Matrix-assisted laser desorption/ionization time-of-flight, *see* MALDI-TOF analysis
Meganuclei, 25; *see also* Macronuclei
Meiosis, 1, 26, 33, 139
   in autogamy, 34
Mendel, 1, 5
Mendelian deletions, 143
Mendelian genetics, 1, 11, 100
Mendelian transmission, 44–45

Metagon hypothesis, 5, 82–86
Mg^{2+} channels, 169
Micronuclei, 26
   activity of, 139–140
   chromosomal breaks, 140–141
   chromosome pairs, 26
   and conjugation, 11, 32
   DNA model, 154, 155
   lack of, 38
   mating type determination, 57
   mitosis, 30
   in *P. aurelia* complex, 26, 29
   ploidy levels, 139
   rDNA genes, 153–154
   telomeres, 140–141
Mirabilis, 64
Mitochondria, 86–88
   and antibiotics, 88, 89
   cell to cell transfer, 88, 90
   character genetics, 88–91
   enzyme variation, 89–90
   and epigenetic inheritance, 175
   functions, 91
   genetic code, 92
   genome in *Paramecium,* 91–92
   interaction with nuclear genome, 90–91
   mutations, 91
   origins, 87
   recombination, 89
   and tetrazolium salt (TTC), 90
Mitosis, 30, 143
Monodidinium, 23
Morphology, 21, 22
MRNA
   and antigen variation, 128
   gene identification, 48, 145
   for immobilization antigens, 117, 119
Mu particles, 76, 77, 83
Mutants, 78–79; *see also* Symbionts
   behavioral, 165
   cell models, 165–166
   Cha (chameleon), 169
   cnr (ciliary nonreversible), 168
   cortical, 158–160
   genetics of, 165
   homology-dependent, 178–179, 181
   induction, 179
   pantophobiac, 168, 169
   paranoiac, 46, 165
   pawn, 168
   pure E mating type, 56–57
   repair, 179

N

Na^+ channels, 169
N-ethylmaleimide-sensitive factor (NSF), 161

Index

Neurospora, 1
Nomenclature
   of immobilization antigens, 102, 103
   for kappa, 68
   for mating types, 13, 16
   other ciliate species, 17
   of *P. aurelia*, 15, 17
   species *versus* varieties, 15–16
   wild-type organisms, 54
Nuclei, 24–29; *see also* Macronuclei; Micronuclei
   changes at conjugation, 31, 33
   dualism, 145–146
   terminology, 25

O

OFAGE, 151
O* phenotype, 56
Ophioglossum reticulatum, 145
Opisthokonta, 20
Orthogonal-field-alternation gel electrophoresis (OFAGE), 151

P

P. aurelia, 5
   life cycle, 36
   move to new location, 16
   nomenclature, 15, 17
   species, 13
   stock, 11
P. aurelia species complex, 16
   evolution, 155–156
   macronuclei, 24–25, 26
   micronuclei, 26
P. avenae, 76
P. biaurelia, 15
   killers, 76, 81
   symbionts in, 79, 80, 81
P. bursaria, 6, 17
   autogamy in, 35
   and chloroplasts, 64
   mating types, 13, 58
P. caudatum, 35, 36
   mating types, 17, 58–59
   nomenclatural problems, 17
   selfers, 59
   symbionts, 81
P. multimicronucleatum, 13
   mating types, 53, 59–60
P. novaurelia, 16
P. octaurelia, 36, 70, 76
P. primaurelia, 4, 15; *see also* Immobilization antigens
   antigen type by stock, 111
   autogamy, 35
   location, 16
   mate killers, 77
   mating types, 52–53
   selfers, 57–58
   wild stocks of, 111
P. quadecaurelia, 16
P. septaurelia, 56, 90
P. sexaurelia, 16
P. sonneborni, 13
P. tetraurelia, 32, 34; *see also* Immobilization antigens
   autogamy, 35
   behavioral mutants, 165
   conjugation, 33
   and gene silencing, 47–48
   genome, 48, 49, 92
   inducing macronuclear regeneration, 36
   killers, 73–76
   mitochondrial genome, 91
   pure E mutant, 56–57
   selfers, 54–55
P. triaurelia, 15, 58
PAK (Paramecium K$^+$) genes, 169, 171
Paramecium aurelia, see *P. aurelia*
Parasites, 99–100
Parasomal sac, 22, 27
Paroral cone, 24, 33
Parthenogenesis, 34
Pawn models, 165
Pawn mutants, 168
Peas, 1, 3
Pelargonium, 64
Phenomic lag, 51
Phenotype
   determination, 11, 30, 36
   O*, 56
Pisum sativum, 1
Plasmagene hypothesis, 5, 64–68
Plasmids, 67, 74–75, 144
Plasmon, 1
Posttranscriptional gene silencing (PTGS), 47, 48, 179–182
Procaryotes, 20
Programmed DNA degradation (PDD) proteins, 182
Prokaryotes, 20
Protein phosphohorylation, 160
Protista
   groups, 20
   origin, 19
Protozoa
   early research, 3–5
   genetics of, 3
   terminology, 19
Pseudocaedibacter conjugatus, 76
Pseudomonas taeniospiralis, 76

PTGS, *see* Posttranscriptional gene silencing (PTGS)

R

Random amplified polymorphic DNA-polymerase chain reaction (RAPD-PCR), 16
RAPD-PCR, *see* Random amplified polymorphic DNA-polymerase chain reaction (RAPD-PCR)
Rapid lysis killers, 77
R bodies, 68, 71, 72, 75, 82; *see also* Kappa
RDNA, *see* Ribosomal DNA (rDNA)
  identification, 48
Recombination, 11, 30, 44, 89
Regulator of chromosome condensation (RCC), 161–162
Rejuvenescence, 2
  and conjugation, 37–38
Reproduction, *see* Asexual fission; Autogamy; Conjugation; Cytogamy; Hemixis
Reversed polarity (RP) mutants, 158
Rhizaria, 20
Ribosomal DNA (rDNA), 151
  clone mapping, 152
  early research, 151–152
  micronuclear genes, 153–154
  structure, 151, 152
  variants, 154, 156
Ribosomal RNA (rRNA)
  early research, 151–152
Ribosomes, 24
Rickettsia, 67, 82
RNA identification, 48
RNA-induced silencing complex (RISC), 182
RNA inhibitor (RNAi), 144, 145
RNA interfering (RNAi), 181
RP mutants, *see* Reversed polarity (RP) mutants
RRNA, *see* Ribosomal RNA (rRNA)

S

Secretory processes, 161
Selfers, 13
  and cytoplasmic state, 54–56
  mating type, 57–58
  in *P. caudatum,* 59
Serotype, 102–103, 115; *see also* Immobilization antigens
  variants, 47
Serotype inheritance, 175, 176
Sexes, 13; *see also* Mating types
Sexual recombination, 44
Soluble NSF attachment receptors (SNARES), 161

Soma, 1
Sonneborn, T.M., 3, 4, 5
Species
  defined, 15
  geographical distribution, 16
  identification, 13, 16
  mating reactions, 13
  and mating types, 12–13
  molecular identification, 16
  nomenclature, 15–16
  versus varieties, 13, 15
Spirostomum, 146
Stenostomum, 157
Stocks, 11, 21, 43–44
  cryopreservation, 7
  intraspecies crosses, 15
Structural inheritance, 160
Stylonychia, 92
Surface proteins, *see* Immobilization antigens
Swimming behavior, 165
  reverse, 166
Symbionts, 92–93; *see also* Kappa; Killer paramecia
  alpha, 79–81
  bacterial affinities, 70
  Caedibibacter caryophila, 82
  Holospora caryophila, 79–81
  lambda, 70, 77, 78
  51ml, 78
  maintenance genes, 82–86, 175
  mate killers, 76–77
  mutants, 78–79
  naming, 68
  nuclear, 79–81
  in *P. caudatum,* 82
  in *P. novaurelia,* 82
  rapid lysis killers, 77
  separation, 78
  sigma, 77
  swimmers, 79
  transmissibility, 70
  types, 69
Symbiotic algae, 24
Syngens, 15, 58; *see also* Species

T

Taxonomy, 16–17; *see also* Nomenclature
Tea (tetraethylammonium), 168
Tectobacter vulgrais, 79
Telomere addition regions (TARs), 179
Telomeres, 140–141, 179
Tetraethylammonium *(Tea),* 168
Tetrahymena, 2, 7
  chromosomal breaks, 141
  mitochondrial DNA, 91
  mitochondrial genetic code, 92

ribosomal RNA and DNA, 151
survival without conjugation, 38
taxonomy, 17
telomerase in, 141
Thermotaxis, 165
Thigmotaxis, 165
Transcription, 128, 129
Trichocysts, 22, 28, 29, 160–162
    defensive function, 23
Trypanosoma, 126
Tubulins, 160

V

Variation, 2
    hereditary verusus environmental, 2, 21
    in mitochondria, 89–90
    during old age, 37
    of rDNA, 154, 156
    serotype, 47
    in single-celled organisms, 20
Varieties
    versus species, 13, 15

W

Watson-Crick theory, 2
Wild-type organisms
    curing mutants, 49
    and gene silencing, 47
    and inheritance patterns, 48
    nomenclature, 54

Y

Yeasts, 21, 66, 86, 88, 92

QL
368
.H87
B43
2008